Instructor's Resource Guide

to accompany

Statistics
A First Course

Sixth Edition

Donald H. Sanders
Educational Consultant

Robert K. Smidt
California Polytechnic State University, San Luis Obispo

Prepared by
John A. Banks
San Jose City College and Evergreen Valley College

Boston Burr Ridge, IL Dubuque, IA Madison, WI New York San Francisco St. Louis
Bangkok Bogotá Caracas Lisbon London Madrid
Mexico City Milan New Delhi Seoul Singapore Sydney Taipei Toronto

McGraw-Hill Higher Education
A Division of The McGraw-Hill Companies

Instructor's Resource Guide to accompany
STATISTICS: A FIRST COURSE, SIXTH EDITION

This book is printed on acid-free paper.

1 2 3 4 5 6 7 8 9 0 QPD QPD 9 0 3 2 1 0 9

ISBN 0-07-229555-4

www.mhhe.com

Contents

Preface

This *Instructor's Resource Guide* is intended to be used with the sixth edition of *Statistics: A First Course* by Donald Sanders and Robert Smidt. It consists of two units.

Unit 1

Unit 1 contains answers to even-numbered Self-Testing Review problems followed by the answers to even-numbered end-of-chapter exercises. Where appropriate, output from MINITAB™ Release 12.1 is placed after the written solution. This enables MINITAB users to compare the two results and check their own MINITAB output. Differences between the hand-worked solutions and the MINITAB solutions are due to rounding and can be ignored.

Unit 2

This unit contains Chapter Examinations for each chapter. Each chapter has three different types of exams:
True/False: Answers are provided in the margin and after the end of the exams.
Multiple Choice: Answers are provided in the margin and after the end of the exams.
Problems and Open-Ended: Sample answers are provided after the end of the exams.
These questions are available as part of the computerized test bank that accompanies the textbook; contact your McGraw-Hill Higher Education sales representative for further information.

Previous editions of this *Instructor's Resource Guide* included transparency masters. In place of these, McGraw-Hill now provides PowerPoint™ presentations based on illustrations from the textbook. PowerPoint files are available for download at www.mhhe.com/sanderssmidt. A hard-copy version of the PowerPoint presentation, which can be used to create overhead transparencies, is available through your McGraw-Hill Higher Education sales representative.

For more information about MINITAB contact:
MINITAB Inc.
3081 Enterprise Drive
State College, PA 16801-3008
www.minitab.com

Unit 1

Answers to Even-Numbered Self-Testing Review and Answers to Even-Numbered End of Chapter Exercises

Chapter 1 Let's Get Started

Answers to Even-Numbered Exercises

2. g

4. e

6. c

8. b

10. Stratified

12. Cluster

14. Simple random

16. Stratified

18. The answer depends on the procedures you've selected.

Chapter 2 Thinking Critically About Data: Liars, #$%& Liars, and a Few Statisticians

Answers to Even-Numbered Exercises

2. Answers will vary. Students afraid of failing will probably lean towards Plummer, while students more confident might choose Pietkowski.

4. The sample is biased, *MacWorld* is a magazine aimed at Macintosh enthusiasts.

6. Yes. For one thing, the study says nothing about other age groups. Also, employment opportunities in Alaska have changed with the discovery of oil. As a result more work age people have moved to Alaska.

8. This press release does not say what kind of average it is (mean, median, or mode) nor does it say anything about the dispersion.

10. 2% is only those who used Grogain and applied for the refund. It does not count those who were not helped by Grogain but did not apply for the refund.

12. No--it's an increase of 100 percent.

14. The statement does not say anything about the kind of teachers in the group. Were they history teachers, math teacher, art teachers, etc. The statement also does not say anything about the kind of history test that was given.

16. Yes. The set-up time has gone from 27.9% to 7.5% while the time the machine runs has increased from 36.6% to 56.7%.

18. Pipe use increases as age increases while the use of "chew or snuff" decreases.

20. 15 percent more 10th graders than 8th graders reported drinking alcohol, and about 10 percent more 12th graders than 10th graders reported drinking alcohol.

22. Answers will vary. The percentage of high school seniors who reported using alcohol or drugs seems to decrease until the early 90's . Since then, use has increased. Most of the increase has been in the use of marijuana. However, over half the high school seniors reportedly used alcohol in the last 30 days.

24. Answers will vary.

26. The LEP students seem to be concentrated in "port of entry states" and border states.

Chapter 3 Descriptive Statistics

Answers to Even-Numbered Self-Testing Review

3-1 Introduction to Data Collection

2. Attribute

4. Numeric, continuous

6. Numeric, discrete

8. Numeric, continuous

10. Attribute

12. Numeric, continuous

3-2 Data Organization and Frequency Distributions

2. The range is 64 − 25 = 39.

4. Yes.

6.

11.18	11.04	7.29	6.35	6.30	6.20	5.71	5.56	5.45	5.35
5.22	5.10	5.01	5.00	4.93	4.90	4.72	4.70	4.68	4.40
4.24	4.20	4.18	4.15	4.14	4.09	4.00	3.95	3.93	3.89
3.73	3.70	3.54	3.43	3.40	3.39	3.29	3.22	3.13	3.05
2.96	2.91	2.89	2.69	2.66	2.57	2.43	2.39	2.37	2.21
2.10	2.01	1.98	1.78	1.63	1.41	1.29	-0.54	-1.36	

8.

Earnings per share ($)	Number of stocks (Frequencies)
−2 < 0	2
0 < 2	5
2 < 4	25
4 < 6	21
6 < 8	4
8 < 10	0
10 < 12	2
	59

10. The range = \$65.50 − \$21.50 = \$44.00.

12. Yes.

14.

25	25	27	32	32	33	33	33	34	34	36	37	38	38	39
40	42	43	43	46	46	46	47	47	48	49	50	51	52	52
52	53	53	54	55	57	64	65	65	65	67	67	75	76	77
78	82	83	83	87	90	91	94	103	106	108	127	130	131	

16.

Price per share ($)	Number of firms (Frequencies)
25 < 40	15
40 < 55	19
55 < 70	8
70 < 85	7
85 < 100	4
100 < 115	3
115 < 130	1
130 < 145	2
	59

18. The range = 309 − 88 = 221.

3-3 Graphic Presentations of Frequency Distributions

2.

Average GMAT Score	Number of Schools (f)
595 < 605	3
605 < 615	4
615 < 625	3
625 < 635	3
635 < 645	9
645 < 655	1
655 < 665	1
665 < 675	0
675 < 685	1
	25

4.

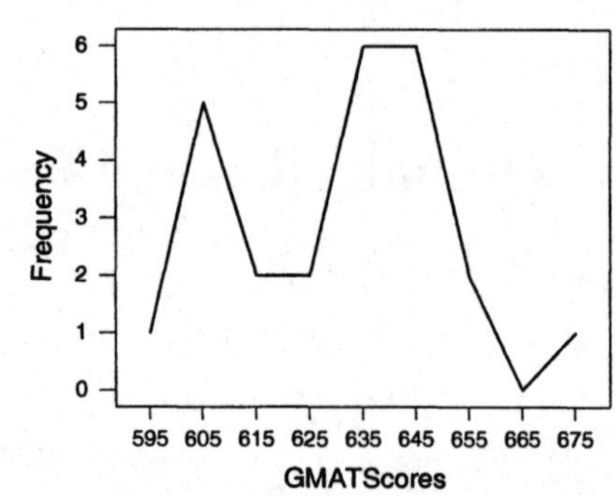

6.

Dotplot for Weeks of Therapy

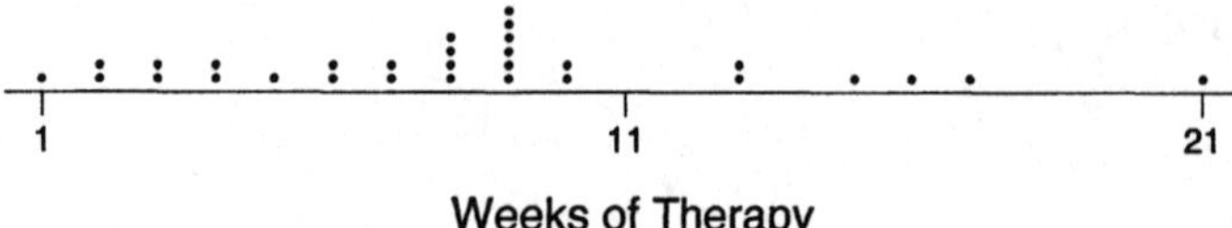

Weeks of Therapy

8.

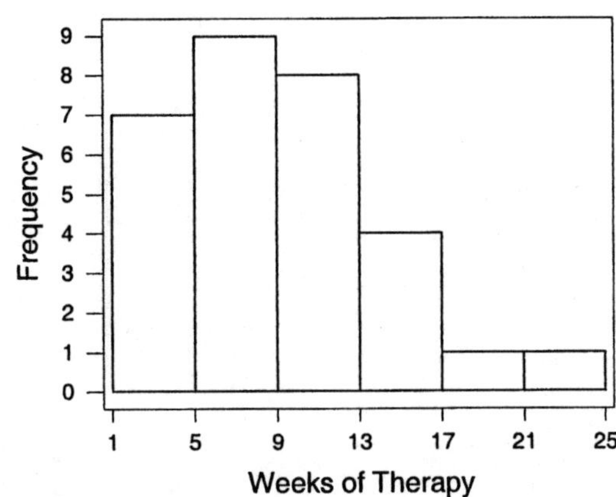

10.

Stem	Leaf
0	1 2 2 3 3 4 4
0	5 6 6 7 7 8 8 8 8 9 9 9 9 9 9
1	0 0 3 3
1	5 6 7
2	1

Key: 0 1 means 1 week

12.

Tax (in \$)	Number of States (f)
1,600 < 2,800	1
2,800 < 4,000	3
4,000 < 5,200	5
5,200 < 6,400	8
6,400 < 7,600	17
7,600 < 8,800	15
8,800 < 10,000	1
10,000 < 11,200	1
	51*

*Includes the District of Columbia.

14.

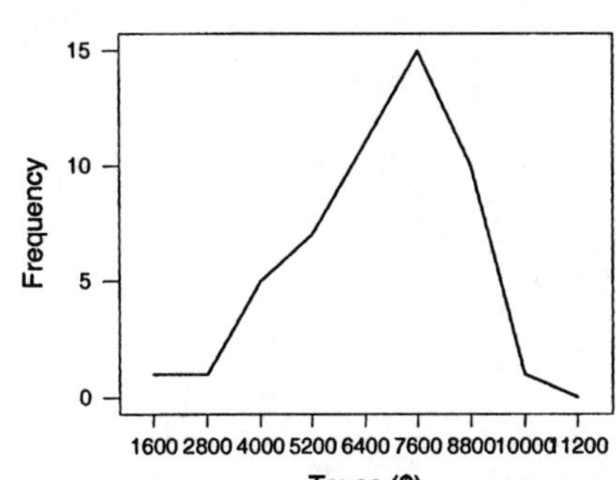

16.

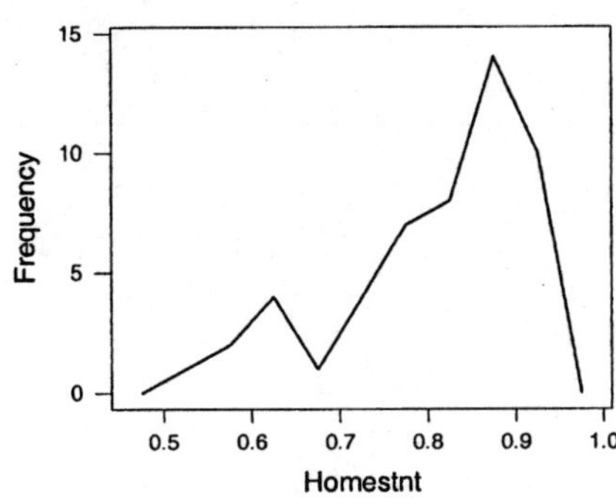

18.

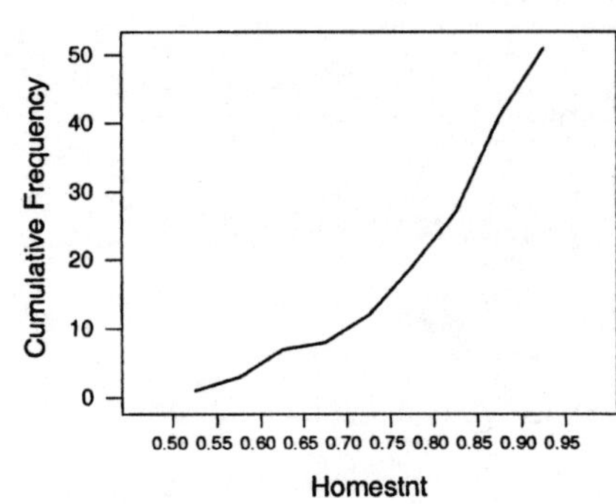

20.

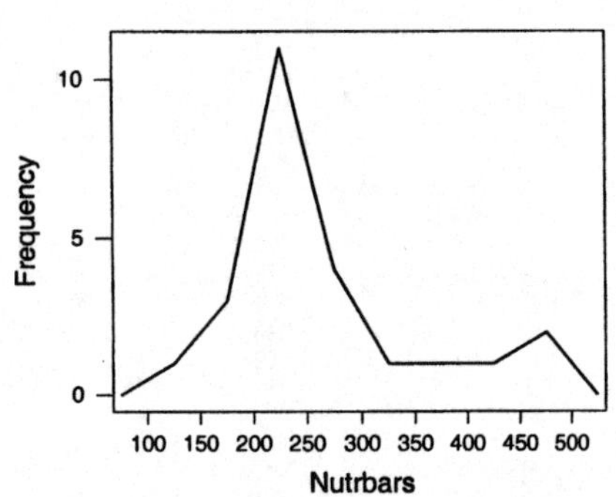

22.

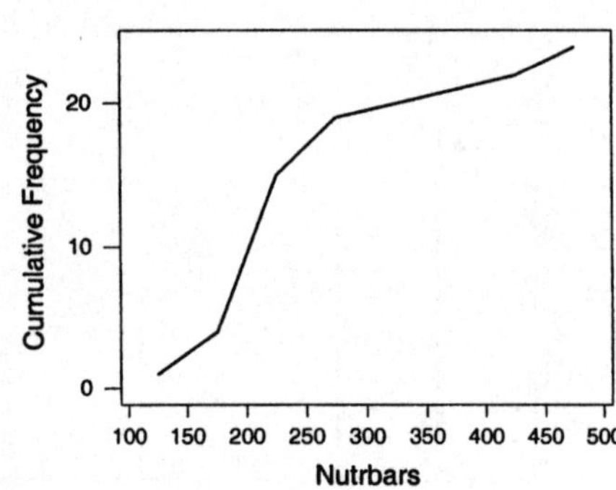

24.

Dotplot for Nutrbars

200 300 400 500

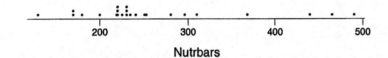

Nutrbars

3-4 Computing Measures of Central Tendency

2. $\overline{x} = 34.67$ assemblies.

4. $\overline{x} = \$42.44$.

6.

Price (in \$)	Number of Stocks (f)
20 < 30	7
30 < 40	2
40 < 50	1
50 < 60	2
60 < 70	5
	17

8. 305 calories.

10. Trimmed Mean $= \dfrac{290 + 280 + 290 + 250}{4} = 277.5$.

12. 5.78 megabytes.

14. $\overline{x} = 541.29$ pins.

16. Two modes: 520 and 535. (Bimodal.)

18.

Bowling Scores	Number of Bowling Teams (f)	Midpoint, m	mf
460 < 480	1	470	470
480 < 500	0	490	0
500 < 520	0	510	0
520 < 540	6	530	3180
540 < 560	4	550	2200
560 < 580	2	570	1140
580 < 600	1	590	590
	14		7580

$\overline{x} = \frac{7580}{14} = 541.4$ pins. (Compare this to the value we found in Exercise 14.)

20. 1.7

22. ~~12~~, ~~13~~, ~~14~~, 14, 15, 16, 16, 16, 16, 18, 18, 20, ~~20~~, ~~20~~, ~~22~~.

So the trimmed mean $= \frac{14+15+16+16+16+16+18+18+20}{9} = \frac{149}{9} = 16.56.$

24. 13.38 13.43 13.88 | 13.94 | 13.98 13.98 14.02. The median is 13.94.

26. ~~13.38~~, 13.43, 13.88, 13.94, 13.98, 13.98, ~~14.02~~

So the trimmed mean $= \frac{13.43+13.88+13.94+13.98+13.98}{7} = \frac{96.61}{7} = 13.801.$

28. 717, 757, 796, 806, 822, 828, 830, 836, 838, 853, 864, 869, 892, 901, 914, 922, 929, 932, 936, 940, 965, 1,003. So the median $= \frac{864+869}{2} = 866.5.$

3-5 Computing Measures of Dispersion and Relative Position

2. The Range = 85 − 1 = 84 pts.

4. $$\sigma^2 = \frac{(50-57.33)^2+(85-57.33)^2+(45-57.33)^2+(60-57.33)^2+(1-57.33)^2+(85-57.33)^2+(50-57.33)^2+(85-57.33)^2+(55-57.33)^2}{9} = 638.$$

6. $$\overline{x} = \frac{1,256+2,726+1,224+2,588+3,294+1,893+2,537+3,177+2,460}{9} = \frac{21155}{9} = 2350.5.$$

8. $s^2 = 561,714.$

10. $z = \frac{x-\overline{x}}{s} = \frac{1256-2350.5}{749.5} = -1.46.$

12. $s = \sqrt{10.2720} = 3.205$ mpg.

14. Using $\overline{x} = 27.455$, we get $z = \frac{x-\overline{x}}{s} = \frac{29-27.455}{3.205} = 0.48.$

16. $$\overline{x} = \frac{7.92+6.94+7.14+6.67+7.12+6.67+7.04+6.66+6.95+5.95}{10} = \frac{69.060}{10} = 6.906.$$

18. $z = \dfrac{5.95 - 6.906}{0.4984} = -1.092.$

20. Since two of the ten scores are below 6.665, 6.665 represents the 20th percentile.

22. Range is $196.7 million.

24. Using $\mu = 110.17$ and $\sigma = 54.61$, $z = \dfrac{x - \mu}{\sigma} = \dfrac{22.7 - 110.17}{54.61} = -1.60.$

26. Since eight of the ten realized tax savings are below 160.75, 160.75 represents the 80th percentile.

28. $\mu \pm 2\sigma$; 52.84 and 75.96 points.

30. $\mu \pm \sigma$; 58.62 and 70.18.

32. $\mu \pm 3\sigma$; 47.06 and 81.74.

34. $\mu \pm 3\sigma$; 46 and 64 degrees.

36. $\mu \pm 2\sigma$; 49 and 61 degrees.

38. $\overline{x} = \dfrac{157}{25} = 6.28$

40.

x	$(x - \overline{x})$	$\lvert x - \overline{x}\rvert$
7	0.72	0.72
6	−0.28	0.28
8	1.72	1.72
5	−1.28	1.28
9	2.72	2.72
6	−0.28	0.28
6	−0.28	0.28
7	0.72	0.72
7	0.72	0.72
4	−2.28	2.28
13	6.72	6.72
2	−4.28	4.28
1	−5.28	5.28
20	13.72	13.72
1	−5.28	5.28
5	−1.28	1.28
7	0.72	0.72
6	−0.28	0.28
7	0.72	0.72
1	−5.28	5.28
8	1.72	1.72
4	−2.28	2.28
8	1.72	1.72
3	−3.28	3.28
6	−0.28	0.28
		63.84

$$\text{MAD} = \frac{\Sigma\lvert x - \overline{x}\rvert}{n} = \frac{63.84}{25} = 2.554.$$

42. $z = \dfrac{x - \overline{x}}{s} = \dfrac{10 - 6.28}{3.974} = 0.936.$

44. 4, 7.5.

46.

Number of years of coaching	Frequency f
$1 < 5$	7
$5 < 9$	15
$9 < 13$	1
$13 < 17$	1
$17 < 21$	1

48.

Deaths, x	x^2
820	672400
751	564001
730	532900
714	509796
759	576081
782	611524
748	559504
720	518400
700	490000
684	467856
732	535824
789	622521
826	682276
773	597529
696	484416
732	535824
715	511225
627	393129
702	492804
687	471969
699	488601
650	422500
16036	11741080

$$\overline{x} = \frac{\Sigma x}{n} = \frac{16036}{22} = 728.91; \; s = \sqrt{\frac{n(\Sigma x^2) - (\Sigma x)^2}{n(n-1)}} = \sqrt{\frac{22(11741080) - (16036)^2}{22(21)}} = 49.90.$$

50. $z = \dfrac{x - \overline{x}}{s} = \dfrac{620 - 728.91}{49.90} = -2.18; \; z = \dfrac{x - \overline{x}}{s} = \dfrac{200 - 141.14}{30.19} = 1.95.$

52. By Chebyshev's theorem at least 75% of the data is between $\overline{x} \pm 2s$, or between 629 and 829 deaths; at least 89% of the data is between $\overline{x} \pm 3s$, or between 579 and 879 deaths.

54. Since there are 38 students, the median is the average of the 19th and 20th values which are both in the 7 to < 10 days absent class

56. Again we are given the size of the first class and number of classes. We make each class the same size and obtain the following frequency distribution:

Sales	Number of Days (f)
$\$750 < \$2,250$	4
$\$2,250 < \$3,750$	17
$\$3,750 < \$5,250$	6
$\$5,250 < \$6,750$	1
$\$6,750 < \$8,250$	1
$\$8,250 < \$9,750$	1
	30

58. Since $0.80(30) = 24$, there are 24 sales less than or equal to the 80th percentile and 6 sales greater than it; thus, the 80th percentile is any value between the 24th value, $\$3,876$, and the 25th value, $\$3,972$. If we average these, one possible value for the 80th percentile is $\$3,924$.

60. We expand the frequency distribution found in Exercise 56, and use $\overline{x} = 3550$ that was found in Exercise 59.

Sales	Number of Days (f)	Class midpoint (m)	Deviation ($m - \overline{x}$)	$(m - \overline{x})^2$	$f(m - \overline{x})^2$
$\$750 < \$2,250$	4	1500	−2050	4202500	16810000
$\$2,250 < \$3,750$	17	3000	−550	302500	5142500
$\$3,750 < \$5,250$	6	4500	950	902500	5415000
$\$5,250 < \$6,750$	1	6000	2450	6002500	6002500
$\$6,750 < \$8,250$	1	7500	3950	15602500	15602500
$\$8,250 < \$9,750$	1	9000	5450	29702500	29702500
	30				78675000

$$s = \sqrt{\frac{\Sigma f(m - \overline{x})^2}{n - 1}} = \sqrt{\frac{78675000}{30 - 1}} = \sqrt{2712931.034} = 1647.10.$$

3-6 Summarizing Qualitative Data

2. Public four-year schools = 10.00%, public two-year schools = 34.52%, private four-year schools = 45.81%, and private two-year schools = 9.68%.

4. Smokers = 13.61%, Nonsmokers = 63.11%.

Answers to Even-Numbered Exercises

2. We are given the size of the first class and number of classes. We make each class the same size and obtain the following frequency distribution:

Price of Scale	Number of Scales
\$10 and less than \$25	15
\$25 and less than \$40	11
\$40 and less than \$55	8
\$55 and less than \$70	3
\$70 and less than \$85	1
\$85 and less than \$100	0
\$100 and less than \$115	0
\$115 and less than \$130	1

4. The ogive is found by finding the cumulative sum of the class of the frequency distribution.

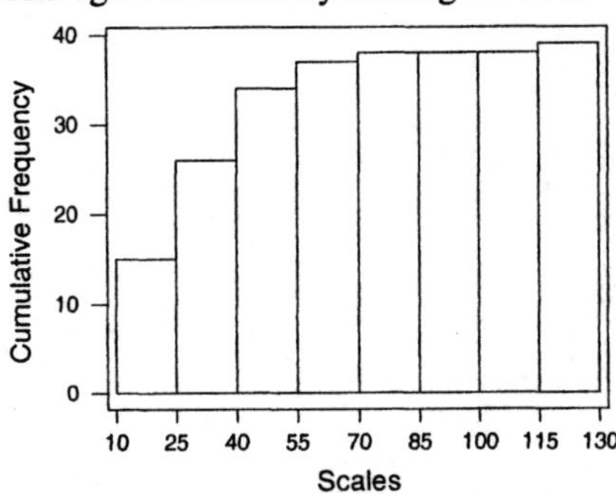

6.

Stem	Leaf
1	0 0 2 2 4 5 7
2	0 0 0 2 2 3 4 4 5 5 5 8
3	0 0 0 2 5 5 5
4	0 8
5	0 0 0 0 0 0
6	0 5 5
7	9
8	
9	
10	
11	
12	0

8. Using the values found in Exercises 5, $Q_1 = \$20$, $Q_3 = \$50$, the interquartile range is $Q_3 - Q_1 = \$50 - \$20 = \$30$. The quartile deviation, QD, is one-half the interquartile range, so $\text{QD} = Q_3 - Q_1 = \dfrac{\$50 - \$20}{2} = \$15$.

10. We expand the frequency distribution found in Exercise 2, and use $\overline{x} = 35.58$ that was found in Exercise 9.

Price of Scale	Number of Scales (f)	Class midpoint (m)	Deviation ($m - \overline{x}$)	$(m - \overline{x})^2$	$f(m - \overline{x})^2$
\$10 and less than \$25	15	17.5	−18.08	326.89	4903.35
\$25 and less than \$40	11	32.5	−3.08	9.49	104.39
\$40 and less than \$55	8	47.5	11.92	142.09	1136.72
\$55 and less than \$70	3	62.5	26.92	724.69	2174.07
\$70 and less than \$85	1	77.5	41.92	1757.29	1757.29
\$85 and less than \$100	0	92.5	56.92	3239.89	0
\$100 and less than \$115	0	107.5	71.92	5172.49	0
\$115 and less than \$130	1	122.5	86.92	7555.09	7555.09
Totals	39				17630.91

$$s = \sqrt{\frac{\Sigma f(m - \overline{x})^2}{n-1}} = \sqrt{\frac{17630.91}{39-1}} = \sqrt{463.97} = 21.54.$$

12. We are given the size of the first class and number of classes. We make each class the same size and obtain the following frequency distribution:

Tuition	Number of Schools
$4,000 and less than $6,000	1
$6,000 and less than $8,000	2
$8,000 and less than $10,000	0
$10,000 and less than $12,000	2
$12,000 and less than $14,000	1
$14,000 and less than $16,000	1
$16,000 and less than $18,000	15
$18,000 and less than $20,000	3

14. Since 20% of 25 is 5, we eliminate the 5 lowest tuition scores and 5 highest tuition *scores*.

$$\bar{x}_{\text{trimmed}} = \frac{\begin{array}{c}13,129 + 15,020 + 16,000 + 16,020 + 16,229 + 16,300 + 16,950 + 17,300 + \\ 17,300 + 17,500 + 17,600 + 17,655 + 17,700 + 17,745 + 17,750\end{array}}{15}$$

$$= \frac{250,198}{15} = 16,679.9$$

16. Since there are 25 tuition values, the 40th percentile is any value between the 10th and 11th tuition, which are $16,229 and $16,300. Taking the average, we get $16,264.50.

18. Using the values found in Exercises 15, $Q_1 = \$14,074.50$, $Q_3 = \$17,747.50$, the interquartile range is $Q_3 - Q_1 = \$17,747.50 - \$14,074.50 = \$3,673$. The quartile deviation is one half the interquartile range, $\text{QD} = Q_3 - Q_1 = \frac{\$17,747.50 - \$14,074.50}{2} = \$1836.50$.

20. We expand the frequency distribution found in Exercise 12, and use $\mu = 15240$ that we found in Exercise 19. This is a population, so we use the population formula.

Tuition	Number of Schools (f)	Class midpoint (m)	Deviation ($m - \mu$)	$(m - \mu)^2$	$f(m - \mu)^2$
$4,000 and less than $6,000	1	5000	−10240	104857600	104857600
$6,000 and less than $8,000	2	7000	−8240	67897600	135795200
$8,000 and less than $10,000	0	9000	0	0	0
$10,000 and less than $12,000	2	11000	−4240	17977600	35955200
$12,000 and less than $14,000	1	13000	−2240	5017600	5017600
$14,000 and less than $16,000	1	15000	240	57600	57600
$16,000 and less than $18,000	15	17000	1760	3097600	46464000
$18,000 and less than $20,000	3	19000	3760	14137600	42412800
	25				370560000

$$\sigma = \sqrt{\frac{\Sigma f(m-\mu)^2}{N}} = \sqrt{\frac{370560000}{25}} = \sqrt{14822400} = 3849.99.$$

22. $$\mu = \frac{\begin{array}{c}17 + 24 + 27 + 29 + 34 + 35 + 40 + 41 + 41 + 44 + 46 + \\ 47 + 47 + 48 + 51 + 60 + 60 + 63 + 65 + 67 + 73\end{array}}{21} = 45.67.$$

24. The position of Q_1 in the array is $\frac{n+1}{4} = \frac{21+1}{4} = 5.5$, so Q_1 is the average of the 5th and 6th positions. Thus $Q_1 = \frac{34+35}{2} = 34.5$. The position of Q_3 is $\frac{3(n+1)}{4} = \frac{3(21+1)}{4} = 16.5$. Since both the 16th and 17th score is 60, $Q_3 = 60$.

26. $$z = \frac{x - \bar{x}}{s} = \frac{50 - 45.67}{15.12} = 0.29.$$

28. $\mu = \dfrac{25.50 + 30.75 + 31.25 + 33.50 + 35.00 + 53.25 + 57.25 + 58.50 + 104.75}{9} = \dfrac{429.75}{9} = \$47.75$

30. The range $= \$104.75 - 25.50 = \$79.25$

32. Using the formula 3.7, $\sigma = \sqrt{\dfrac{\Sigma(x-\mu)^2}{N}}$ and $\mu = \$47.75$ from Exercise 28.

Price per share, x	$(x-\mu)$	$(x-\mu)^2$
25.50	−22.25	495.0625
30.75	−17.00	289.0000
31.25	−16.50	272.2500
33.50	−14.25	203.0625
35.00	−12.75	162.5625
53.25	5.50	30.2500
57.25	9.50	90.2500
58.50	10.75	115.5625
104.75	57.00	3249.0000
		4907.0000

So, $\sigma = \sqrt{\dfrac{4907.0000}{9}} = \sqrt{545.222} = 23.35.$

34.

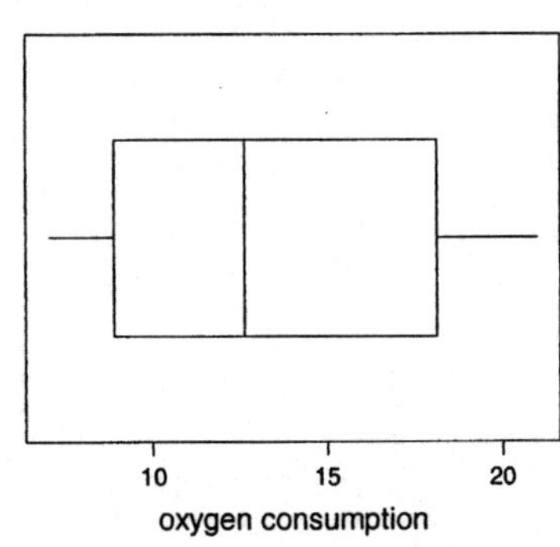

36. Since 6.55 is between the 7th and 8th rating score, 6.55 is the 70th percentile.

38. $z = \dfrac{x-\mu}{\sigma} = \dfrac{5.60 - 6.386}{0.6778} = -1.16.$

40. Expand the frequency distribution to obtain the table:

Grades	Frequency, f	Midpoint, m	fm
40 < 50	4	45	180
50 < 60	6	55	330
60 < 70	10	65	650
70 < 80	4	75	300
80 < 90	4	85	340
90 < 100	2	95	190
	30		1990

So $\mu = \dfrac{1990}{30} = 66.33.$

42.

Stem	Leaf
4	1 2 7 9
5	0 2 5 6 7 9
6	0 1 3 5 5 5 6 8 8 9
7	2 4 8 9
8	1 5 7 8
9	0 5

44. The range $= 3,465 - 2,448 = 1017$.

46. $\overline{x} = \dfrac{2,829 + 2,359 + 1,784 + 1,569 + 1,517 + 1,301 + 1,267 + 873 + 750 + 655}{10}$
$= \dfrac{14,904}{10} = 1,490.4.$

48. The third of the ten employees is 873 and the fourth is 1,267, so the 30th percentile is $\dfrac{873 + 1267}{2} = \dfrac{2140}{2} = 1070.$

50. $z = \dfrac{x - \overline{x}}{s} = \dfrac{1,267 - 1,490}{695} = -0.32.$

52. $\overline{x} = \dfrac{387 + 440 + 567 + 615}{4} = \dfrac{2009}{4} = 502.25.$

54. Construct the table using $\overline{x} = 502.25$.

Calories	$\lvert x - \overline{x}\rvert$
387	115.25
440	62.25
567	64.75
615	112.75
	355.00

$\text{MAD} = \dfrac{\Sigma\lvert x - \overline{x}\rvert}{n} = \dfrac{355.00}{4} = 88.75.$

56. Construct the table using $\overline{x} = 502.25$.

Calories	$(x - \overline{x})$	$(x - \overline{x})^2$
387	−115.25	13282.5625
440	−62.25	3875.0625
567	64.75	4192.5625
615	112.75	12712.5625
		34062.7500

$s = \sqrt{\dfrac{\Sigma(x - \overline{x})^2}{n - 1}} = \sqrt{\dfrac{34062.75}{4 - 1}} = \sqrt{11354.25} = 106.56$ calories.

58. The 20th percentile is 80.5°.

60. The median temperature $= \dfrac{82 + 84}{2} = 83$ degrees.

62. Using Chebyshev's Theorem with $k = 3$, we know that at least 89% of functional families should score between $\mu - 3\sigma$ and $\mu + 3\sigma$ scale points. $\mu - 3\sigma = 7.17 - 3(1.49) = 2.70$ and $\mu + 3\sigma = 7.17 + 3(2.49) = 11.64$.

64. Using Chebyshev's Theorem with $k = 3$, we know that at least 89% of dysfunctional families should score between $\mu - 3\sigma$ and $\mu + 3\sigma$ scale points. $\mu - 3\sigma = 5.57 - 3(2.49) = -1.90$ and $\mu + 3\sigma = 5.57 + 3(2.49) = 13.04$.

66.

Test scores	Students (f)	Midpoint (m)	fm
40 < 50	10	45	450
50 < 60	5	55	275
60 < 70	7	65	455
70 < 80	3	75	225
80 < 90	16	85	1360
90 < 100	12	95	1140
	53		3905

$$\mu = \frac{3905}{53} = 83.7.$$

68. $\overline{x} = \dfrac{5+8+9+10+16}{5} = \dfrac{48}{5} = 9.6$ tickets.

70.

Stem	Leaf
1	2 4 5 9
2	2 3 5
3	7
4	1 3
5	1 4 5
6	2 5 6 8
7	1 3
8	4 5 8
9	2 2 6

72. Range $= 1403 - 269 = 1134$.

74. Using $\overline{x} = 860.2$, compute the table:

Exam Scores, x	$\|x - \overline{x}\|$
269	591.2
345	515.2
381	479.2
428	432.2
455	405.2
582	278.2
760	100.2
763	97.2
898	37.8
990	129.8
999	138.8
1059	198.8
1141	280.8
1181	320.8
1213	352.8
1264	403.8
1352	491.8
1403	542.8
15483	5796.6

$$\text{MAD} = \frac{\Sigma|x - \overline{x}|}{n} = \frac{5796.6}{18} = 322.03.$$

76. $\overline{x} = \dfrac{15,536 + 14,193 + 13,316 + 11,740 + 13,733 + 15,077 + 15,704}{7} = \dfrac{99,299}{7} = 14,185.6.$

78. Using $\overline{x} = 14,185.6$ from exercise 76.

Number of applications, x	$x - \overline{x}$	$(x - \overline{x})^2$
15,536	1350.4	1823580.16
14,193	7.4	54.76
13,316	−869.6	756204.16
11,740	−2445.6	5980959.36
13,733	−452.6	204846.76
15,077	891.4	794593.96
15,704	1518.4	2305538.56
99,299		11865777.72

$$s = \sqrt{\frac{\Sigma(x - \overline{x})^2}{n - 1}} = \sqrt{\frac{11865777.72}{6}} = \sqrt{1977629.62} = 1406.3.$$

80. $z = \dfrac{x - \overline{x}}{s} = \dfrac{8.00 - 7.37}{3.54} = 0.18.$

82. $z = \dfrac{x - \overline{x}}{s} = \dfrac{30 - 26.67}{14.07} = 0.24.$

Chapter 4 Probability Concepts

Answers to Even-Numbered Self-Testing Review

4-1 Some Basic Considerations

2. Since there are 12 face cards out of 52, the probability is $\frac{12}{52} = \frac{3}{13}$ or 0.2308.

4. $\frac{74 \text{ thousand}}{152 \text{ thousand}} = \frac{37}{76} = 0.4868.$

6. 0.68

8. $\frac{183}{2347} = 0.0780.$

10. No, the probability of an event cannot be more than 1.

12. No, since a probability can't be negative.

14. $\frac{255}{2470} = 0.1032.$

16. $\frac{8,018}{79,618} = 0.1007.$

4-2 Probabilities for Compound Events

2. $\frac{1}{10}.$

4. This is an "and" situation and the events are independent, so we multiply the individual probabilities: $\frac{1}{10} \cdot \frac{1}{10} \cdot \frac{1}{10} = \frac{1}{1000} = 0.001.$

6. "At least a 4" means a 4 or a 5 or a 6 in this problem. Since the events are mutually exclusive (a roll of a 4 on a single trial excludes a roll of a 5 on that same trial) we can add $\frac{1}{6} + \frac{1}{6} + \frac{1}{6} = \frac{3}{6} = 0.5.$

8. These events are mutually exclusive, so we add $\frac{1}{6} + \frac{3}{6} = \frac{4}{6} = 0.6667.$

10. $\frac{2}{10} \cdot \frac{2}{10} = \frac{4}{100} = 0.04.$

12. $(0.2) \cdot (0.8) = 0.16.$

14. Using the rule for complements, $1 - 0.64 = 0.36.$

16. $\frac{2}{10} \cdot \frac{1}{9} = \frac{2}{90} = 0.0222.$

18. $\frac{2}{10} \cdot \frac{8}{9} = \frac{16}{90} = 0.1778.$

20. Since the probability of the complement (both items are good) is $\frac{56}{90} = 0.6222$, we'll use the rule for complements to compute the probability that at least one is defective. That is, $1 - 0.6222 = 0.3778$.

22. $\frac{1683}{2825} = 0.5958$.

24. $\frac{262}{2825} + \frac{126}{2825} = \frac{388}{2825} = 0.1373$.

26. $\frac{3+8+13}{50} = \frac{24}{50} = 0.48$.

28. $\frac{13+23+8}{130} = \frac{44}{130} = 0.3385$.

30. $\frac{178}{199} = 0.8945$.

32. $\frac{75}{178} = 0.4213$.

34. $1 - \frac{8}{178} = \frac{170}{178} = 0.9551$.

36. $\frac{28+14}{60} = \frac{42}{60} = 0.7$.

38. $\frac{30+25}{100} = \frac{55}{100} = 0.55$.

40. $\frac{55}{100} + \frac{50}{100} - \frac{25}{100} = \frac{80}{100} = 0.8$.

42. No, since $P(\text{saying "no" and female}) = \frac{25}{100} \neq 0$.

44. $\frac{34}{122} = 0.2787$.

46. $\frac{19}{122} = 0.1557$.

48. $1 - \frac{40}{122} = \frac{82}{122} = 0.6721$.

50. $\frac{19}{43} = 0.4419$.

52. $\frac{19}{34} = 0.5588$.

54. No. $P(\text{female}) = \frac{79}{122} = 0.6475$; $P(\text{female}|\text{sprint}) = \frac{15}{34} = 0.4412$; $P(\text{female}) \neq P(\text{female}|\text{sprint})$.

56. $1 - \frac{9736}{12002} = 0.1888$.

58. $\frac{544}{12002} + \frac{6860}{12002} - \frac{322}{12002} = \frac{7082}{12002} = 0.5901$.

60. Events are not mutually exclusive since being female does not exclude one from being from Philomath. There were 4,259 who were females and from Philomath.

62. $\frac{5936}{79,619} = 0.0746$.

64. $\frac{50,474}{79,619} = 0.6339$.

66. $\frac{9,539}{79,619} = 0.1198$.

68. $\frac{14,571}{79,619} + \frac{50,474}{79,619} - \frac{9,539}{79,619} = \frac{55,506}{79,619} = 0.6971$.

70. $\frac{12,047}{50,474} + \frac{9,454}{50,474} = \frac{21,501}{50,474} = 0.4260$.

72. $1 - \frac{233}{14,571} = \frac{14,338}{14,571} = 0.9840$.

4-3 Random Variables, Probability Distributions, and Expected Values

2. $E(x) = 0.174 + 0.678 + 0.705 + 0.764 + 0.085 + 0.054 + 0.063 + 0.153 + 0.09 = 2.766.$

4. 0.14.

6.

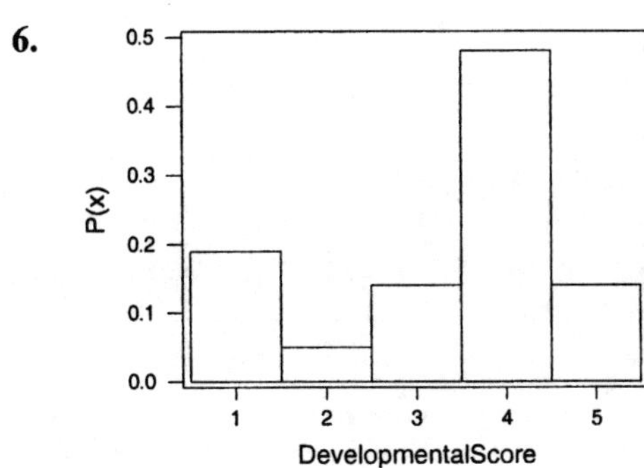

8. $\Sigma[x^2P(x)] = 1^2(0.19) + 2^2(0.05) + 3^2(0.14) + 4^2(0.48) + 5^2(0.14) = 12.83.$
$\sigma = \sqrt{\Sigma[x^2P(x)] - \mu^2} = \sqrt{12.83 - 3.33^2} = \sqrt{1.7411} = 1.320.$

10.

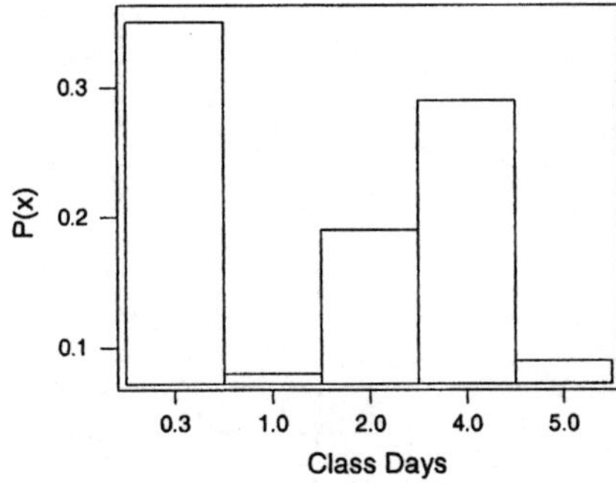

12. $E(x) = 1(0.08) + 2(0.19) + 3(0.35) + 4(0.29) + 5(0.09) = 3.12.$

14. $E(x) = 0.60\ (\$20,000) + 0.40\ (-\$25,000) = \$2,000.$ She should make the investment

16. Add the probability of selling 3 cars or 4 cars to get: $0.15 + 0.05 = 0.20.$

18. $E(x) = 0(0.20) + 1(0.30) + 2(0.30) + 3(0.15) + 4(0.05) = 1.55.$

20. $\frac{44}{59} = 0.7458.$

22. $1 - \left(\frac{27}{59} + \frac{11}{59}\right) = 1 - \frac{38}{59} = \frac{21}{59} = 0.3559.$

24. $$E(x) = \frac{0(44) + 3(1) + 4(4) + 6(1) + 7(1) + 8(2) + 9(1) + 11(1) + 12(1) + 14(1) + 16(1) + 19(1)}{59}$$
$$= \frac{129}{59} = 2.19.$$

26.

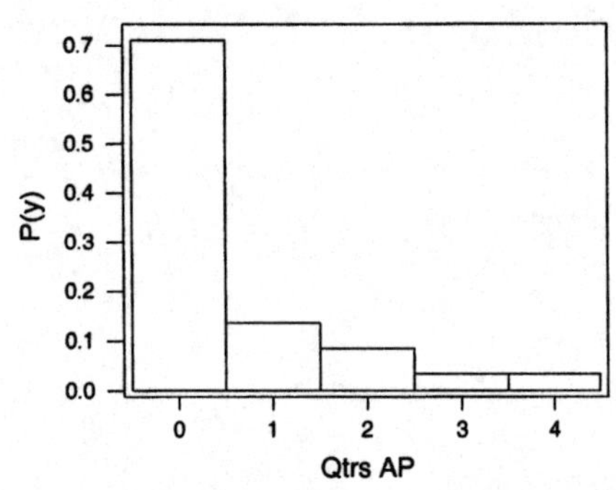

28. $\Sigma[y^2P(y)] = 0^2\left(\frac{42}{59}\right) + 1^2\left(\frac{8}{59}\right) + 2^2\left(\frac{5}{59}\right) + 3^2\left(\frac{2}{59}\right) + 4^2\left(\frac{2}{59}\right) = \frac{78}{59}.$

$\sigma = \sqrt{\Sigma[y^2P(y)] - \mu^2} = \sqrt{\frac{78}{59} - \left(\frac{32}{59}\right)^2} = \sqrt{1.027866} = 1.0138.$

30. $E(w) = 0\left(\frac{27}{59}\right) + 1\left(\frac{11}{59}\right) + 2\left(\frac{6}{59}\right) + 3\left(\frac{2}{59}\right) + 4\left(\frac{6}{59}\right) + 6\left(\frac{2}{59}\right) + 9\left(\frac{2}{59}\right) + 10\left(\frac{2}{59}\right) + 12\left(\frac{1}{59}\right)$

$= \frac{115}{59} = 1.95.$

32. $\frac{3}{30} = 0.10.$

34. $P(11 \text{ or} 12 \text{ or } 14 \text{ or } 15 \text{ or } 22 \text{ assist}) = \frac{5}{30} + \frac{4}{30} + \frac{2}{30} + \frac{1}{30} + \frac{1}{30} = \frac{13}{30} = 0.4333.$

36.

x	$P(x)$	$x \cdot P(x)$	$x^2 \cdot P(x)$
4	0.066667	0.26667	1.0667
5	0.100000	0.50000	2.5000
6	0.133333	0.80000	4.8000
7	0.133333	0.93333	6.5333
8	0.166667	1.33333	10.6667
9	0.100000	0.90000	8.1000
10	0.066667	0.66667	6.6667
11	0.100000	1.10000	12.1000
12	0.033333	0.40000	4.8000
13	0.033333	0.43333	5.6333
14	0.066667	0.93333	13.0667
		8.2667	75.933

$\sigma = \sqrt{\Sigma[x^2P(x)] - \mu^2} = \sqrt{75.933 - (8.2667)^2} = \sqrt{7.5947} = 2.7558.$

38.

y	$P(y)$	$y \cdot P(y)$	$y^2 \cdot P(y)$
4	0.066667	0.26667	1.0667
5	0.133333	0.66667	3.3333
6	0.033333	0.20000	1.2000
7	0.100000	0.70000	4.9000
8	0.100000	0.80000	6.4000
9	0.066667	0.60000	5.4000
10	0.066667	0.66667	6.6667
11	0.166667	1.83333	20.1667
12	0.133333	1.60000	19.2000
14	0.066667	0.93333	13.0667
15	0.033333	0.50000	7.5000
22	0.033333	0.73333	16.1333
		9.5	105.03

$\sigma = \sqrt{\Sigma[y^2 P(y)] - \mu^2} = \sqrt{105.03 - 9.5^2} = \sqrt{14.78} = 3.8444.$

Answers to Even-Numbered Exercises

2. Using the relative frequency formula this probability is $\dfrac{16,715}{25,519} = 0.6550.$

4. Since there is replacement of the first card before selection of the second card we use the formula $P(\text{ace}_1 \text{ and king}_2) = P(\text{ace}_1) \cdot P(\text{king}_2) = \dfrac{4}{52} \cdot \dfrac{4}{52} = \dfrac{16}{2704} = 0.0059.$

6. $P(\text{ace and king}) = P(\text{ace}_1 \text{ and king}_2) + P(\text{king}_1 \text{ and ace}_2) = 0.0059 + 0.0059 = 0.0118.$

8. Since the first card is not replaced before the second card is selected we use the formula $P(\text{ace}_1 \text{ and king}_2) = P(\text{ace}_1) \cdot P(\text{king}_2 \mid \text{ace}_1) = \dfrac{4}{52} \cdot \dfrac{4}{51} = \dfrac{16}{2652} = 0.0060.$

10.
$$\begin{aligned} P(\text{ace and king}) &= P(\text{ace}_1 \text{ and king}_2) + P(\text{king}_1 \text{ and ace}_2) \\ &= P(\text{ace}_1) \cdot P(\text{king}_2 \mid \text{ace}_1) + P(\text{king}_1) \cdot P(\text{ace}_2 \mid \text{king}_1) \\ &= \frac{4}{52} \cdot \frac{4}{51} + \frac{4}{52} \cdot \frac{4}{51} = \frac{16}{2652} + \frac{16}{2652} \\ &= 0.0060 + 0.0060 = 0.0120. \end{aligned}$$

12. $P(\text{def}_1 \text{ and def}_2) = P(\text{def}_1) \cdot P(\text{def}_2 \mid \text{def}_1) = \dfrac{7}{26} \cdot \dfrac{6}{25} = \dfrac{42}{650} = 0.0646.$

14. $P(\text{at least one def}) = 1 - P(\text{both not def}) = 1 - 0.5262 = 0.4738.$

16. $P(\text{at least a 4}) = P(4) + P(5) + P(6) = 0.25 + 0.20 + 0.15 = 0.60.$

18. $P(> 34) = P(35 - 44) + P(45 \text{ or more}) = 0.316 + 0.106 = 0.422.$

20. $P(\text{interest in IB}) = \dfrac{99}{431} = 0.2297.$

22. $P(\text{female and interest in IB}) = \dfrac{46}{431} = 0.1067.$

24. $P(\text{woman}) = \dfrac{9}{50} = 0.18.$

26. $$P(\text{woman or pop. vote}) = P(\text{woman}) + P(\text{pop. vote}) - P(\text{woman and pop. vote})$$
$$= \frac{9}{50} + \frac{18}{50} - \frac{8}{50} = \frac{19}{50} = 0.38.$$

28. $P(\text{3 bdr.}) = \dfrac{425}{1,706} = 0.2491.$

30. $$P(\text{Center City or 3 bdr.}) = P(\text{Center City}) + P(\text{3 bdr.}) - P(\text{Center City and 3 bdr.})$$
$$= \frac{533}{1,706} + \frac{425}{1,706} - \frac{97}{1,706} = \frac{861}{1,706} = 0.5047.$$

32. $P(\text{3 bdr.} \mid \text{Center City}) = \dfrac{97}{533} = 0.1820.$

34. No these events are not independent since $P(\text{Center City}) \neq P(\text{Center City} \mid \text{3 bdr.})$.

36. $P(\text{Brooks County}) = \dfrac{3,388}{16,715} = 0.2027.$

38. $P(\text{community college and Brooks County}) = \dfrac{1,073}{16,715} = 0.0642.$

40. $$P(\text{com. col. or Dover County}) = P(\text{com. col.}) + P(\text{Dover County}) - P(\text{com. col. and Dover County})$$
$$= \frac{3,458}{16,715} + \frac{2,515}{16,715} - \frac{618}{16,715} = \frac{5,355}{16,715} = 0.3204.$$

42. $P(\text{Morgan County} \mid \text{4yr col.}) = \dfrac{3,021}{12,155} = 0.2485.$

44. $P(\text{not commercial}) = 1 - P(\text{commercial}) = 1 - \dfrac{62}{268} = \dfrac{206}{268} = 0.7687.$

46. $P(\text{Highly polluted} \mid \text{urban}) = \dfrac{73}{90} = 0.8111.$

48. $$P(\text{Rural or not highly pol.}) = P(\text{Rural}) + P(\text{Low pol.}) + P(\text{mod. pol.})$$
$$- [P(\text{Rural and low pol.}) + P(\text{Rural and mod. pol.})]$$
$$= \frac{65}{268} + \frac{51}{268} + \frac{67}{268} - \left[\frac{33}{268} + \frac{23}{268}\right] = \frac{127}{268} = 0.4739.$$

50. $$P(\text{Rural and not highly pol.}) = P(\text{Rural and low pol.}) + P(\text{Rural and mod. pol.})$$
$$= \frac{33}{268} + \frac{23}{268} = \frac{56}{268} = 0.2090.$$

52. Continuous.

54. Discrete.

56. Discrete.

58. Continuous.

60. Discrete.

62. Continuous.

64.

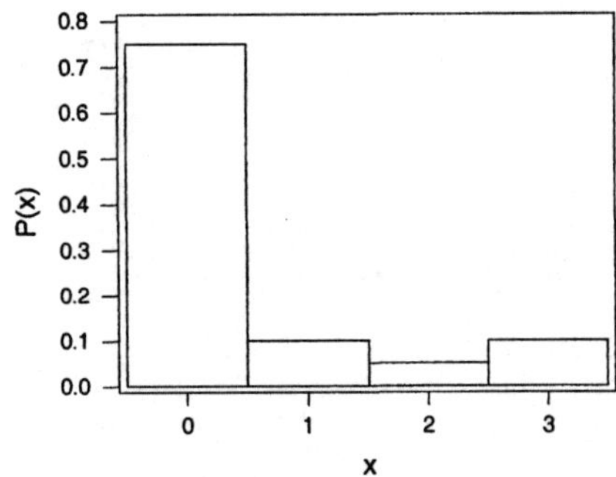

66.

x	$P(x)$	$x \cdot P(x)$	$x^2 \cdot P(x)$
0	0.75	0	0
1	0.10	0.10	0.10
2	0.05	0.10	0.20
3	0.10	0.30	0.90
		0.50	1.20

$\sigma = \sqrt{\Sigma[x^2 P(x)] - \mu^2} = \sqrt{1.20 - (0.50)^2}$
$= \sqrt{0.95} = 0.9747.$

68.

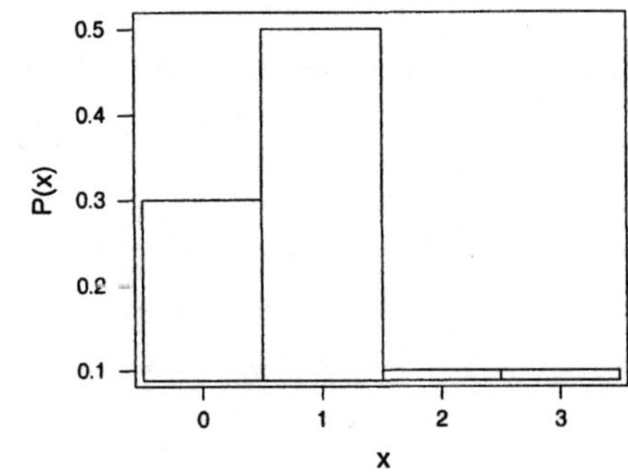

70.

x	$P(x)$	$x \cdot P(x)$	$x^2 \cdot P(x)$
0	0.3	0	0
1	0.5	0.5	0.5
2	0.1	0.2	0.4
3	0.1	0.3	0.9
		1.0	1.8

$\sigma = \sqrt{\Sigma[x^2 P(x)] - \mu^2} = \sqrt{1.8 - (1.0)^2}$
$= \sqrt{0.8} = 0.8944.$

72. Since there are six slots marked less than \$1,000, $P(x < \$1000) = \frac{6}{9} = 0.6667$.

74. $E = -10,000 \cdot 0.40 + 20,000 \cdot 0.50 + 30,000 \cdot 0.10 = \$9,000$.

Chapter 5 Probability Distributions

Answers to Even-Numbered Self-Testing Review

5-1 Binomial Experiments

2. This is not a binomial experiment since there are four (not two) possible outcomes.

4. This is a binomial experiment. Files of 6 liver recipients are checked. A success means the patient is still alive, $n = 6$, $p = 0.83$, $q = 0.17$, and $r = 0,\ 1,\ 2,\ \ldots,\ 6$.

5-2 Determining Binomial Probabilities

2. If at least 10 have been cured, this translates to 10, 11, or 12 that have been cured. Since we must look under $p = 0.20$, we must equivalently determine the probability of 2, 1, or 0 "non-cures." Add $0.0687 + 0.2062 + 0.2835$ to get 0.5584.

4. We translate this into between 3 and 6 (inclusive) have been non-cures, and then find that the probabilities under $p = 0.20$ add to 0.4378.

6. Add the probabilities for $r = 0, 1, 2,$ or 3 to get 1.000. It's practically certain that less than four use marijuana weekly.

8. Since the probability of 0.67 is not found in Appendix 1, this probability must be computed by formula 5.2. For the probability that exactly 10 had income over $\$65,000$:
${}_nC_r p^r q^{n-r} = {}_{12}C_{10}(0.67)^{10}(0.33)^2 = 66(0.67)^{10}(0.33)^2 = 0.1310.$

10. In Exercise 9 we found the probability that at least 10 had an income over $\$65,000$ was 0.1876. Using the compliment rule we get $1 - 0.1876 = 0.8124$.

12. With $n = 20$ and $p = 0.45$, add the probabilities for $r = 10,\ 11,\ 12$ to get $0.1623 + 0.1771 + 0.1593 = 0.4987$.

14. ${}_{14}C_{10}(0.46)^{10}(0.54)^4 = 1001(0.46)^{10}(0.54)^4 = 0.0361.$

16. Using the result of Exercises 14 and 15, we get: $P(r = 10) + P(r = 11) = 0.0361 + 0.0112 = 0.0473$.

18. With $n = 20$ and $p = 0.20$, add the probabilities for $r = 0,\ 1,\ 2,\ 3,\ 4$ to get $0.0115 + 0.0576 + 0.1369 + 0.2054 + 0.2182 = 0.6296$.

20. $E(x) = np = 12 \cdot 0.80 = 9.6$ patients.

22. $E(x) = np = 350 \cdot 0.47 = 164.5$.

24. $E(x) = np = 50 \cdot 0.01 = 0.5.$

26. $E(x) = np = 1,000 \cdot 0.064 = 64.$

28. With $n = 3$ and $p = 0.40$, we get $P(r = 1) = 0.4320$.

30. With $n = 25$ and $p = 0.40$, we get $P(r = 4) = 0.0071$.

32. $E(x) = np = 25 \cdot 0.40 = 10.$

34. With $n = 16$ and $p = 0.15$, we get $P(r = 2) = 0.2775$.

36. With $n = 16$ and $p = 0.15$, we get $P(r = 0 \text{ or } 1) = 0.0743 + 0.2097 = 0.2840$.

38. $\sigma = \sqrt{npq} = \sqrt{16 \cdot 0.15 \cdot 0.85} = \sqrt{2.04} = 1.4283.$

5-3 The Poisson Distribution

2. Using appendix 10, $P(x = 0) = 0.3012$.

4. With $\mu = 20$ and $x = 0, 1, 2, \ldots, 10$, the answer $= 0.0108$.

6. With $\mu = 3$ and $x = 5$, the probability is 0.1008.

8. With $\mu = 2$ and $x = 4$, the probability is 0.0902.

10. With $\mu = 4$ and $x = 6$, the probability is 0.1042.

12. With $\mu = 3$ and $x = 0$, the probability is 0.0498.

14. With $\mu = 10$ and $x = 12$, the probability is 0.0948.

16. With $\mu = 10$ and $x = 16, 17, \ldots, 24$, the probability is 0.0488.

18. With $\mu = 2.5$ and $x = 2$, the probability is 0.2565.

20. With $\mu = 2.5$ and $x = 0, 1$, the probability is 0.2873.

22. With $\mu = 6$ and $x = 5$, the probability is 0.1606.

24. With $\mu = 6$ and $x = 0, 1, 2, 3$, the probability is 0.1512.

26. With $\mu = 10$ and $x = 8, 9$, the probability is 0.2377.

28. With $\mu = 5$ and $x = 0$, the probability is 0.0067.

30. With $\mu = 5$ and $x = 2, 3, \ldots, 15$, we use the complement rule with $x = 0, 1$. The probability is $1 - 0.0404 = 0.9596$.

32. With $\mu = 1$ and $x = 0$, the probability is 0.3679.

34. With $\mu = 1$ and $x = 1, 2, \ldots, 7$, we use the complement rule with $x = 0$. The probability is $1 - 0.3679 = 0.6321$.

5-4 The Normal Distribution

2. $0.4977 - 0.1736 = 0.3241$.

4. $0.5 + 0.4982 = 0.9982$.

6. $0.4838 - 0.4744 = 0.0094$.

8. $0.5 + 0.4591 = 0.9591$.

10. $0.5 + 0.1664 = 0.6664$.

12. $0.5 - 0.3869 = 0.1131$.

14. $0.5 + 0.4931 = 0.9931$.

16. $0.3944 + 0.4951 = 0.8895$.

18. $0.0987 + 0.4970 = 0.5957$.

20. $0.5 - 0.1628 = 0.3372$.

22. $0.4812 + 0.3413 = 0.8225$.

24. $0.5 - 0.4452 = 0.0548$.

26. $0.5 - 0.3413 = 0.1587$.

28. 0.0861.

30. 0.0139.

32. $0.4861 - 0.4452 = 0.0409$.

34. $0.5 - 0.2881 = 0.2119$.

36. 0.6517.

38. 0.4154.

40. 0.0721.

42. If 0.90 or 90 percent of the area lies to the right of the unknown z value, then 0.40 lies between the mean and the required z score. The area closest to 0.40 found in Appendix 2 is 0.3997. This corresponds in the table to a z score of -1.28.

44. Since $0.50 - 0.25 = 0.25$, and the closest area to 0.25 is 0.2486, the z score is -0.67.

46. This is equivalent to saying 0.4950 or 49.50 percent of the area is between the unknown z score and the mean. The two table entries of 0.4949 and 0.4951 are equally distant from 0.4950 so we use $z = \pm 2.575$.

48. The required z score is -1.645. So $x = 3.5 - 1.645(0.4) = 2.84$ years or about 34 months.

50. $z_1 = \dfrac{74 - 80}{20} = -0.30$ and $z_2 = \dfrac{84 - 80}{20} = 0.20$. So, $P(74 \leq x \leq 84) = P(-0.30 \leq z \leq 0.20) = 0.1179 + 0.0793 = 0.1972$.

52. $z_1 = \dfrac{5500 - 6000}{240} = -2.08$ and $z_2 = \dfrac{6500 - 6000}{240} = 2.08$. So, $P(5500 \leq x \leq 6500) = P(-2.08 \leq z \leq 2.08) = 0.4812 + 0.4812 = 0.9624$.

54. $z = \dfrac{6750 - 6000}{240} = 3.125$. So, $P(x \leq 6750) = P(z \leq 3.125) = 0.5000 + 0.4997 = 0.9997$.

56. The required z score is -0.84. The value that 20% of the compressive strength will be below is $x = 6000 + 240(-0.84) = 5798.4$.

58. The required z score is 1.04. The value that 15% of the compressive strength will be above is $x = 6000 + 240(1.04) = 6249.6$.

60. $z = \dfrac{16 - 12.4}{4} = 0.90$. So, $P(x \leq 16) = P(z \leq 0.90) = 0.5000 + 0.3159 = 0.8159$.

62. The required z score is -0.84. The value that 80% of the seedlings will grow above is $x = 12.4 + 4(-0.84) = 9.04$ inches.

Answers to Even-Numbered Exercises

2. $P(r \geq 5) = 1 - P(r \leq 4) = 1 - P(r = 0) + P(r = 1) + P(r = 2) + P(r = 3) + P(r = 4)$
$= 1 - (0.0037 + 0.0259 + 0.0836 + 0.1651 + 0.2222) = 1 - 0.5005 = 0.4995$.

4. $P(r \geq 1) = 1 - P(r = 0) = 1 - 0.0467 = 0.9533$.

6. Using Appendix 10 with $\mu = 3.9$ we obtain $P(x = 3) = 0.2001$.

8. Using Appendix 10 with $\mu = 6.7$ we obtain $P(x = 0) = 0.0012$.

10. Using Appendix 10 with $\mu = 6.7$ we obtain
$P(x > 16) = P(x = 17, 18, 19) = 0.0004 + 0.0001 + 0.0000 = 0.0005$.

12. Using Appendix 2,
$P(-2.45 \leq z \leq 1.91) = P(-2.45 \leq z \leq 0) + P(0 \leq z \leq 1.91) = 0.4929 + 0.4719 = 0.9648$.
Note: $P(-2.45 \leq z \leq 0) = P(0 \leq z \leq 2.45)$.

14. Using Appendix 2,
$P(-3.05 \leq z \leq -1.73) = P(-3.05 \leq z \leq 0) - P(-1.73 \leq z \leq 0) = 0.4989 - 0.4968 = 0.0021$.

16. Using Appendix 2, $P(z \leq -1.64) = 0.5 - P(-1.64 \leq z \leq 0) = 0.5 - 0.4495 = 0.0505$.

18. Binomial distribution applies with $n = 15$, $r > 8$, and $p = 0.80$. Since $p > 0.50$, to use the table in Appendix 1 we consider the equivalent probability with $n = 15$, $q = 1 - p = 0.20$, and $r \leq 6$;
$P(r = 0, 1, 2, 3, 4, 5, 6) = 0.0352 + 0.1319 + 0.2309 + 0.2501 + 0.1876 + 0.1032 + 0.0430$
$= 0.9819$.

20. Consider the equivalent probability with $n = 20$, $q = 0.40$, and $5 \leq r \leq 8$; so
$P(5 \leq r \leq 8) = 0.0746 + 0.1244 + 0.1659 + 0.1797 = 0.5446$.

22. Using Appendix 10 with $\mu = 5$ we obtain $P(x = 3) = 0.1404$.

24. Using Appendix 10 with $\mu = 5$ we obtain $P(x = 3, 4) = 0.1404 + 0.1755 = 0.3159$.

26. Binomial distribution applies with $n = 7$, $r < 5$, and $p = 0.80$. The equivalent probability with $n = 7$, $q = 0.20$, and $r > 2$; $P(r > 2) = 0.1147 + 0.0287 + 0.0043 + 0.0004 + 0.0000 = 0.1481$.

28. Binomial distribution applies with $n = 17$, $r < 3$, and $p = 0.15$, so
$P(r < 3) = 0.0631 + 0.1893 + 0.2673 = 0.5197$.

30. Binomial distribution applies with $n = 10$, $r < 4$, and $p = 0.60$. The equivalent probability with $n = 10$, q $= 0.40$, and $r > 6$. So $P(r > 6) = 0.0425 + 0.0106 + 0.0016 + 0.0001 = 0.0548$.

32. Binomial distribution applies with $n = 28$ and $p = 0.43$, so $E(x) = np = 28 \cdot 0.43 = 12.04$.

34. Binomial distribution applies with $n = 9$, $r > 5$, and $p = 0.40$ so
$P(r > 5) = 0.0743 + 0.0212 + 0.0035 + 0.0003 = 0.0993$.

36. Binomial distribution applies with $n = 9$, $r < 3$, and $p = 0.40$ so
$P(r < 3) = 0.0101 + 0.0605 + 0.1612 = 0.2318$.

38. Using Appendix 10 with $\mu = 2.4$ we obtain $P(x = 2) = 0.2613$.

40. Binomial distribution applies with $n = 18$, $9 \le r \le 14$, and $p = 0.55$; the equivalent probability is when $n = 18$, $q = 0.45$, and $4 \le r \le 9$. Thus,
$P(4 \le r \le 9) = 0.0291 + 0.0666 + 0.1181 + 0.1657 + 0.1864 + 0.1694 = 0.7353$.

42. Using $\mu = 26$ and s $= 6$, $z = \dfrac{15 - 26}{6} = -1.83$. Thus
$P(x < 15) = P(z < -1.83) = 0.5 - P(-1.83 < z < 0) = 0.5 - 0.4664 = 0.0336$.

44. Binomial distribution applies with $n = 10$, $r = 10$, and $p = 0.65$; the equivalent probability is when $n = 10$, $q = 0.35$, and $r = 0$. Thus $P(r = 0) = 0.0135$.

46. This is a Poisson probability distribution with $\mu = 4.60$ and $x = 0$. Using Appendix 10
$P(x = 0) = 0.0101$.

48. Binomial distribution applies with $n = 20$, $r > 12$, and $p = 0.70$; the equivalent probability is when $n = 20$, $q = 0.30$, and $r < 8$. Thus,
$P(r \le 7) = 0.0008 + 0.0068 + 0.0278 + 0.0718 + 0.1304 + 0.1789 + 0.1916 + 0.1643 = 0.7724$.

50. Binomial distribution applies with $n = 20$, $10 \le r \le 15$, and $p = 0.70$; the equivalent probability is when $n = 20$, $q = 0.30$, and $5 \le r \le 10$. Thus
$P(5 \le r \le 10) = 0.1789 + 0.1916 + 0.1643 + 0.1144 + 0.0654 + 0.0308 = 0.7454$.

52. Using $\mu = 12$ and $\sigma = 2$ in the formula $z = \dfrac{x - \mu}{\sigma}$ to get $z = \dfrac{14 - 12}{2} = 1.00$. Thus
$P(x < 14) = P(z < 1.00) = 0.5 + P(0 < z < 1.00) = 0.5 + 0.3413 = 0.8413$.

54. Using $\mu = 12$ and $\sigma = 2$ in the formula $z = \dfrac{x - \mu}{\sigma}$ to get: $z_1 = \dfrac{600 - 675}{75} = -1.00$ when $x_1 = 600$ and $z_2 = \dfrac{700 - 675}{75} = 0.33$ when $x_2 = 700$. Thus

$$P(600 < x < 700) = P(-1.00 < z < 0.33) = P(-1.00 < z < 0) + P(0 < z < 0.33) \\ = 0.3413 + 0.1293 = 0.4706.$$

56. Binomial distribution applies with $n = 10$, $r > 5$, and $p = 0.30$. Thus
$P(r > 5) = 0.0368 + 0.0090 + 0.0014 + 0.0001 = 0.0473$.

58. Using $\mu = 30.6$ and $\sigma = 9.1$ in the formula $z = \dfrac{x - \mu}{\sigma}$ to get $z = \dfrac{21 - 30.6}{9.1} = -1.05$. Thus
$P(21 < x) = P(-1.05 < z) = P(-1.05 < z < 0) + 0.5 = 0.3531 + 0.5 = 0.8531$.

60. Using $\mu = 23.3$ and $\sigma = 7.7$ in the formula $z = \dfrac{x - \mu}{\sigma}$ to get $z = \dfrac{10 - 23.3}{7.7} = -1.73$. Thus
$P(x < 10) = P(z < -1.73) = 0.5 - P(-1.73 < z < 0) = 0.5 - 0.4582 = 0.0418$.

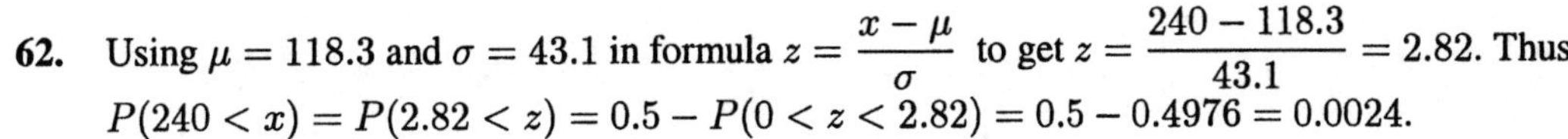

62. Using $\mu = 118.3$ and $\sigma = 43.1$ in formula $z = \dfrac{x-\mu}{\sigma}$ to get $z = \dfrac{240 - 118.3}{43.1} = 2.82$. Thus $P(240 < x) = P(2.82 < z) = 0.5 - P(0 < z < 2.82) = 0.5 - 0.4976 = 0.0024$.

64. Binomial distribution applies with $n = 250$ and $p = 0.39$, so $E(x) = np = 250 \cdot 0.39 = 97.5$.

66. Binomial distribution applies with $n = 530$ and $p = 0.26$, so $E(x) = np = 530 \cdot 0.26 = 137.8$.

68. Here we want to find z_0 so that $P(z < z_0) = 0.30$. Since $P(z < z_0) = 0.5 - P(z_0 < z < 0) = 0.30$, implies $P(z_0 < z < 0) = 0.20$. Looking in Appendix 2 for 0.20 we find $P(0 < z < 0.52) = 0.1985$ and $P(0 < z < 0.53) = 0.2019$. Since 0.1985 is closer to 0.20 than 0.2019 we chose $z_0 = -0.52$.
Note z_0 is negative since $z_0 < z < 0$.

70. Here we want to find z_0 so that $P(-z_0 < z < z_0) = 0.90$. This is equivalent to finding z_0 so that $P(0 < z < z_0) = 0.45$. Looking in Appendix 2 for 0.45 we find $P(0 < z < 1.64) = 0.4495$ and $P(0 < z < 1.65) = 0.4505$. Since these are the same distance from 0.45, thus $z_0 = 1.645$

72. In Exercise 69 we found that $P(z_0 < z) = 0.10$ implies $z_0 = 1.28$. Substituting $\mu = 550$ and $\sigma = 95$ into the formula $z = \dfrac{x-\mu}{\sigma}$ and solving for x yields: $1.28 = \dfrac{x - 550}{95} \quad \Rightarrow \quad x = (1.28)(95) + 550 = 671.6$.

Chapter 6 Sampling Concepts

Answers to Even-Numbered Self-Testing Review

6-1 Sampling: The Need and The Advantages

2. When a population is finite, all the values can be listed. An infinite population has an unlimited number of values.

4. Sampling reduces cost and time, and in some cases is just as accurate as census taking. In the case of an infinite population, sampling must be carried out.

6. All Psychology majors at your school.

8. All Geo Prisms assembled this week.

10. All schizophrenic patients undergoing experimental treatment.

12. All 367 freshmen at Casestudy College.

14. 512

16. False. Statistics taken from samples will most probably differ from population parameter values.

6-2 Sampling Distribution of Means--A Pattern of Behavior

2. A sample distribution consists of n values that have been selected from the population. The sampling distribution consists of the means from all the samples of a given size.

4. The finite population correction factor is needed. Since the cards aren't replaced after they are selected, there are a finite number of possible samples. That is, ${}_{10}C_5 = 252$ samples.

6. The finite population correction factor isn't needed. When sampling is done with replacement, the population is infinite.

8. Finite population correction factor $= 0.8947$, $\sigma_{\bar{x}} = 0.2686$.

10. Finite population correction factor $= 0.9835$, $\sigma_{\bar{x}} = 0.7233$.

12. Finite population correction factor $= 0.9987$, $\sigma_{\bar{x}} = 2.322$.

14. We don't have a sample situation here. Because we are dealing with an individual member of a population, this problem is solved just like those in Chapter 5. That is, we first compute a z value corresponding to a value of 2.08: $z = \frac{x-\mu}{\sigma} = \frac{2.08-2.0}{0.1} = 0.8$. Thus, the probability that one sack weighs more than 2.08 pounds is equal to the probability z is more than 0.8. This is computed as $0.5 - 0.2881 = 0.2119$.

16. The mean of the sampling distribution, being equal to the population mean, is 0.45.

18. Since n is over 30, the sampling distribution is approximately normal.

20. The range is $0.45 \pm (2.00)(0.003)$, or 0.444 to 0.456 inches.

22. If the sample size is 49, the standard error is $\frac{15}{\sqrt{49}} = 2.14$, and the range is $200 \pm 2\ (2.14)$ or 195.72 to 204.28.

24. As the sample size increases, the dispersion of the sampling distribution decreases and the sample mean is likely to have a value closer to the population mean.

26. The standard error is $\frac{30}{\sqrt{36}} = 5$. The z score is $\frac{535-540}{5} = -1.00$, which corresponds to a z table value of 0.3413. Adding $0.3413 + 0.5000$ gives a probability of 0.8413.

28. $z = \frac{17.5-16.9}{3.2/\sqrt{50}} = 1.33$, which yields a table value of 0.4082. And $0.5000 - 0.4082$ gives a probability of 0.0918.

30. $z = \frac{110-100}{\left(20/\sqrt{30}\right)\left(\sqrt{170/199}\right)} = \frac{10}{3.3749} = 2.96$ which yields a table value of 0.4985. The probability that the mean IQ of the sample of 30 exceeds 110 is thus $0.5000 - 0.4985 = 0.0015$.

32. From Exercise 31, we have $\mu_{\bar{x}} = \$64,500$ and $\sigma_{\bar{x}} = \$1,367.36$, so $z = \frac{65,000-64,500}{1367.36} = 0.37$. 0.37 yields a table value of 0.1443. The probability that the mean of the sample will be less than \$65,000 is $0.5000 + 0.1443 = 0.6443$.

6-3 Sampling Distribution of Percentages

2. $\sigma_p = \sqrt{\frac{57 \cdot 43}{10}} = 15.66$.

4. $\sigma_p = \sqrt{\frac{57 \cdot 43}{1000}} = 1.57$.

6. $\sigma_p = \sqrt{\frac{10 \cdot 90}{100}} = 3$.

8. $\sigma_p = \sqrt{\frac{50 \cdot 50}{100}} = 5$.

10. $\sigma_p = \sqrt{\frac{90 \cdot 10}{100}} = 3$.

12. The mean of the sampling distribution is 60 percent because it is equal to the population percentage. $\sigma_p = \sqrt{\frac{60 \cdot 40}{35}} = 8.28.$

14. The standard error is $\sqrt{\frac{8.5 \cdot 91.5}{1,000}} = 0.88$ percent. We want $z\,\sigma_p = 1$ percent. Solving we get $z = \frac{1}{0.88} = 1.13$. The area between the mean and $z = 1.13$ is 0.3708, so there's a $2 \cdot 0.3708$ or a 0.7416 chance the sample percent will be within one percent of π.

16. The $\sigma_p = \sqrt{\frac{64 \cdot 36}{100}} = 4.8$ percent. Thus, $z_1 = \frac{60 - 64}{4.8} = -0.83$, which yields a table value of 0.2967. And $z_2 = \frac{68 - 64}{4.8} = +0.83$, which also gives a table value of 0.2967. Thus, $0.2967 + 0.2967 =$ the probability of 0.5934.

18. The $\sigma_p = \sqrt{\frac{53 \cdot 47}{50}} = 7.06$ percent. Phil loses the election if he receives less than 50 percent of the votes--that is, he loses if the percentage of votes cast by the sample of 50 eligible voters falls into the area below 50 percent under a normal curve that has a population percentage of 53 percent. This area has a z value of $\frac{50 - 53}{7.06} = -0.42$. We see from Appendix 2 that a z value of 0.42 corresponds to an area figure of 0.1628. Thus, the probability that Phil loses is $0.5000 - 0.1628$, or 0.3372.

20. $z = \frac{30 - 29}{\sqrt{\frac{29 \cdot 71}{80}}} = \frac{1}{5.0732} = 0.20$. The table value of 0.20 is 0.0793. The probability that less than 30% are smoker is $0.5000 + 0.0793 = 0.5793$, or almost 58%.

22. $z_1 = \frac{25 - 29}{\sqrt{\frac{29 \cdot 71}{80}}} = \frac{-4}{5.0732} = -0.79$ and $z_2 = \frac{35 - 29}{\sqrt{\frac{29 \cdot 71}{80}}} = \frac{6}{5.0732} = 1.18$. The table value of -0.79 is 0.2852 and the table value of 1.18 is 0.3810. The probability that between 25% and 35% are smoker is $0.2852 + 0.3810 = 0.6662$.

Answers to Even-Numbered Exercises

2. $\sigma_{\bar{x}} = \frac{0.7}{\sqrt{36}} = \frac{0.7}{6} = 0.1167$, so $z = \frac{4.7 - 4.5}{0.1167} = 1.71$;
$P(4.7 < \bar{x}) = 1 - P(0 < z < 1.71) = 0.5000 - 0.4564 = 0.0436.$

4. As n increases, $\sigma_{\bar{x}} = \frac{\sigma}{\sqrt{n}}$ decreases; so the corresponding z value increases and $P(4.7 < \bar{x})$ decreases.

6. $\sigma_{\bar{x}} = \frac{3.72}{\sqrt{20}} = 0.83$, so $z = \frac{9 - 10.43}{0.83} = -1.72$. Thus
$P(9 < \bar{x}) = 1 - P(-1.72 < z < 0) = 0.5000 - 0.4573 = 0.0427.$

8. As n increases, $\sigma_{\overline{x}} = \dfrac{\sigma}{\sqrt{n}}$ decreases; so the corresponding z value increases and $P(9 < \overline{x})$ decreases.

10. $\dfrac{0.683}{2} = 0.3415$ and 0.3415 corresponds to a z value of 1. Thus the sample mean has a 68.3 percent chance of falling between 26 ± 0.6 mg., or 25.4 to 26.6 mg.

12. $\mu_{\overline{x}} = \mu = 12$ days and using Formula 6.4 $\sigma_{\overline{x}} = \dfrac{2}{\sqrt{15}} = 0.52$ days.

14. 95.4 percent corresponds to a z value of 2. Now using this value with the standard deviation of the sampling distribution from Exercise 12, the sample mean has a 95.4 percent chance of falling between $12 \pm 2(0.52)$ days, or 10.96 to 13.04 days.

16. 95.4 percent corresponds to a z value of 2; $\sigma_{\overline{x}} = \dfrac{10}{\sqrt{100}} = 1.00$; so $132 \pm 2(1.00)$ or 130.00 to 134.00 degrees.

18. As the sample size increases $\sigma_{\overline{x}}$ decreases so that the dispersion is concentrated about the mean.

20. 99.7 percent corresponds to a z value of 3 and $\sigma_{\overline{x}} = \dfrac{5}{\sqrt{49}} = 0.714$; so $24 \pm 3(0.714)$; or 21.858 to 26.142.

22. As the sample size increases $\sigma_{\overline{x}}$ decreases so that the dispersion is concentrated about the mean.

24. Here we are asked to find $P(|\overline{x} - \mu_{\overline{x}}| < 0.3)$. So $\sigma_{\overline{x}} = \dfrac{1.02}{\sqrt{64}} = 0.1275$ and $z = \pm\dfrac{0.3}{0.1275} = \pm 2.35$.
Thus $P(|\overline{x} - \mu_{\overline{x}}| < 0.3) = 2 \cdot P(0 < z < 2.35) = 2 \cdot 0.4906 = 0.9812$.

26. Here we are asked to find $P(600 < \overline{x} < 650)$. Calculating $\sigma_{\overline{x}}$ and then the two z values we obtain:
$\sigma_{\overline{x}} = \dfrac{80}{\sqrt{40}} = 12.65$ and $z_1 = \dfrac{600 - 631}{12.65} = -2.45$ and $z_2 = \dfrac{650 - 631}{12.65} = 1.50$. So
$P(600 < \overline{x} < 650) = P(-2.45 < z < 1.50) = 0.4929 + 0.4332 = 0.9261$.

28. $\sigma_p = \sqrt{\dfrac{0.30 \cdot 0.70}{46}} = 0.0676 = 6.75$ percent and $z = \pm\dfrac{0.05}{0.0676} = 0.74$;
probability $= 2 \cdot 0.2704 = 0.5408$.

30. $\sigma_p = \sqrt{\dfrac{0.45 \cdot 0.55}{35}} = 0.0841$, $z_1 = \dfrac{0.40 - 0.45}{0.0841} = -0.59$ and $z_2 = \dfrac{0.50 - 0.45}{0.0841} = 0.59$;
probability $= 2 \cdot 0.2224 = 0.4448$.

32. $\sigma_{\overline{x}} = \dfrac{\$1{,}451}{\sqrt{67}} = \$177.27$; $z = \dfrac{3{,}500 - 3{,}222}{177.27} = 1.57$ probability $= 0.5000 - 0.4418 = 0.0582$.

34. $\sigma_{\overline{x}} = \dfrac{2}{\sqrt{17}} = 0.485$; $z_1 = \dfrac{11 - 12}{0.485} = -2.06$ and $z_2 = \dfrac{13 - 12}{0.485} = 2.06$;
probability $= 2 \cdot 0.4803 = 0.9606$.

36. $\sigma_{\bar{x}} = \dfrac{11}{\sqrt{31}} = 1.976$; $z_1 = \dfrac{25 - 23.3}{1.976} = 0.86$ and $z_2 = \dfrac{27 - 23.3}{1.976} = 1.87$;
probability $= 0.4693 - 0.3051 = 0.1642$.

38. 99.7 percent corresponds to a z value of 3 and $\sigma_{\bar{x}} = \dfrac{57.3}{\sqrt{46}} = 8.45$; $118.3 \pm 3(8.45)$ minute; 92.95 to 143.65 minutes.

Chapter 7 Estimating Parameters

Answers to Even-Numbered Self-Testing Review

7-1 Estimate, Estimation, Estimator, et Cetera

2. f

4. g

6. d

8. On the average, such an estimate will be correct.

10. b

7-2 Interval Estimation of the Population Mean: Some Basic Concepts

2. Using normal distribution probabilities, we know our mad statistician would find about 68.26 percent or 6,826 within $1\sigma_{\bar{x}}$, 95.44 percent or 9,544 within $2\sigma_{\bar{x}}$, and 99.74 percent or 9,974 within $3\sigma_{\bar{x}}$.

4. 90 percent of the sample means will fall within 1.645 standard errors of $\mu_{\bar{x}}$.

6. As the level of confidence increases, the number of standard errors for the interval estimate increases.

8. 92 percent of the sample means will fall within 1.75 standard errors of $\mu_{\bar{x}}$.

10. It defines the probability that the interval will include the true population mean.

12. The 90 percent level results in a smaller maximum error of estimate.

14. At a given level of confidence, the error bound is the maximum distance between the point estimate and the true population mean. Since this distance could be to the left or to the right of the point estimate, the confidence interval width is twice the value of the error bound.

16. The level of confidence is 90 percent. The parameter is the population mean (which is unknown). The point estimate is 84.1 (the known sample mean) and the error bound is 5.3.

7-3 Estimating the Population Mean

2. Half of 0.96 is 0.48. The area closest to 0.48 in the table is 0.4798, which corresponds to $z = 2.05$. So 96 percent of the area under the normal curve lies between vertical lines erected at $z = -2.05$ and $z = +2.05$.

4. To read the values for the t distributions, we must first calculate the corresponding area under the curve found to the right of a vertical line at the required t value. Since the level of confidence of 90 percent specifies the area in the center of the curve, this implies 100 – 90 percent or 10 percent remains for the areas in the two tails. To find the area for the right tail, take half of 10 percent or 5 percent. This tells us to look down the column headed 0.05 in the table in Appendix 4. Since $n = 12$ and $\text{df} = n - 1$, we look across the 11 df line to obtain the t value of 1.796. Thus, 90 percent of the area in this t distribution lies between vertical lines erected at ± 1.796.

6. The area in the right tail is $\frac{100-95}{2}$ or 0.025 and $\text{df} = 14$, so $t = \pm 2.145$. You might find it helpful to write the corresponding level of confidence in the t distribution table in your text above the values listed for the right tails. That is, a level of confidence of 80 percent corresponds to the area of 0.10 in the right tail, a level of confidence of 90 percent corresponds to an area of 0.05 in the right tail, and so on. This way you do not have to "reinvent the wheel" for each small sample confidence level problem.

8. Although the sample is small, the population is normally distributed and the σ is known. Thus, we use the normal distribution with $z = \pm 1.645$.

10. For a small sample with σ unknown, we use a t distribution. With 16 df, the required t values are ± 1.746.

12. The estimated standard error of the mean is $\frac{12}{\sqrt{58}} = 1.576$.

14. The maximum error of estimate is $z_{\alpha/2}\hat{\sigma}_{\bar{x}} = 1.96 \cdot 1.576 = 3.088$.

16. The point estimate is the sample mean of 24.7.

18. Use $t_{0.005} = 2.898$ since this is a small sample and σ is unknown.

20. The 99 percent confidence interval is 24.7 ± 3.620 or 21.080 to 28.320.

22. The estimated standard error of the mean is $\frac{3.81}{\sqrt{17}} = 0.924$.

24. The maximum error of estimate is $z_{\alpha/2}\hat{\sigma}_{\bar{x}} = 1.96 \cdot 0.924 = 1.811$.

26. The area in the right tail is $\frac{100-95}{2}$ or 0.025 and $\text{df} = 20 - 1 = 19$, so $t_{0.025} = \pm 2.093$. The estimated standard error is $\frac{0.7}{\sqrt{20}} = 0.157$. So the 95 percent confidence interval is $2.3 \pm (2.093)(0.157)$, which is 1.971 to 2.629 hours per night.

28. The estimated standard error is $\frac{5}{\sqrt{15}} = 1.291$. Since $\text{df} = 14$, $t_{0.05} = 1.761$ and the confidence interval is $17.6 \pm (1.761)(1.291)$ or 15.327 to 19.873.

30. The sample mean is $\frac{128}{7} = 18.286$ hours, the sample standard deviation is 5.529 hours, $\text{df} = 6$ and $t_{0.025} = 2.447$. The estimated standard error is $\frac{5.529}{\sqrt{7}} = 2.09$. The confidence interval is $18.286 \pm 2.447 \cdot 2.09$ or 13.172 to 23.400 hours.

32. The estimated standard error is $\frac{4.8}{\sqrt{71}} = 0.570$. The confidence interval is $30.1 \pm 1.645 \cdot 0.570$ or 29.162 to 31.038.

34. The estimated standard error is $\frac{58}{\sqrt{126}} = 5.167$. The confidence interval is $648 \pm 1.645(5.167)$ or 639.5 to 656.5 hours.

36. We calculate the mean using the first column, then use it to calculate the other two columns.

x	$x - \overline{x}$	$(x - \overline{x})^2$
490	226.93	51497.2249
123	−140.07	19619.6049
320	56.93	3241.0249
120	−143.07	20469.0249
290	26.93	725.2249
270	6.93	48.0249
300	36.93	1363.8249
68	−195.07	38052.3049
280	16.93	286.6249
230	−33.07	1093.6249
530	266.93	71251.6249
450	186.93	34942.8249
130	−133.07	17707.6249
250	−13.07	170.8249
95	−168.07	28247.5249
3946		288716.9335

$\overline{x} = \frac{3946}{15} = 263.07$. Then $s = \sqrt{\frac{\Sigma(x - \overline{x})^2}{n-1}} = \sqrt{\frac{288716.9335}{14}} = 143.61$. Since df $= 14$, $t_{0.05} = \pm 1.761$. The estimated standard error is $\widehat{\sigma}_{\overline{x}} = \frac{s}{\sqrt{n}} = \frac{143.61}{\sqrt{15}} = 37.08$.

$$\overline{x} - t_{0.05}\,\widehat{\sigma}_{\overline{x}} < \mu < \overline{x} + t_{0.05}\,\widehat{\sigma}_{\overline{x}}$$
$$263.07 - (1.761)(37.08) < \mu < 263.07 + (1.761)(37.08)$$
$$263.07 - 65.30 < \mu < 263.07 + 65.30$$
$$197.77 < \mu < 328.37.$$

38. We calculate the mean using the first column, then use it to calculate the other two columns.

x	$x - \overline{x}$	$(x - \overline{x})^2$
2000	−493.33	243374.4889
2050	−443.33	196541.4889
2129	−364.33	132736.3489
2200	−293.33	86042.4889
2500	6.67	44.4889
2500	6.67	44.4889
2666	172.67	29815.9289
2800	306.67	94046.4889
3595	1101.67	1213676.7899
22440		1996326.1100

$\overline{x} = \dfrac{22440}{9} = 2493.33$. Then $s = \sqrt{\dfrac{\Sigma(x-\overline{x})^2}{n-1}} = \sqrt{\dfrac{1996326.1100}{8}} = 499.54$. Since df $= 8$, $t_{0.05} = \pm 1.860$. Since this is a finite population we use $\widehat{\sigma}_{\overline{x}} = \dfrac{s}{\sqrt{n}}\sqrt{\dfrac{N-n}{N-1}} = \dfrac{499.54}{\sqrt{9}}\sqrt{\dfrac{53-9}{53-1}} = 153.17$.
Thus the confidence interval is

$$\overline{x} - t_{0.05}\,\widehat{\sigma}_{\overline{x}} < \mu < \overline{x} + t_{0.05}\,\widehat{\sigma}_{\overline{x}}$$
$$2493.33 - (1.860)(153.17) < \mu < 2493.33 + (1.860)(153.17)$$
$$2493.33 - 284.90 < \mu < 2493.33 + 284.90$$
$$\$2208.43 < \mu < \$2778.23.$$

40. We calculate the mean using the first column, then use it to calculate the other two columns.

x	$x-\overline{x}$	$(x-\overline{x})^2$
225	36.37	1322.7769
194	5.37	28.8369
191	2.37	5.6169
173	−15.63	244.2969
176	−12.63	159.5169
225	36.37	1322.7769
155	−33.63	1130.9769
170	−18.63	347.0769
1509		4561.8752

$\overline{x} = \dfrac{1509}{8} = 188.63$. Then $s = \sqrt{\dfrac{\Sigma(x-\overline{x})^2}{n-1}} = \sqrt{\dfrac{4561.8752}{7}} = 25.53$. Since df $= 7$, $t_{0.005} = \pm 3.499$.
Since this is a finite population, so we use the finite population correction factor:
$\widehat{\sigma}_{\overline{x}} = \dfrac{s}{\sqrt{n}}\sqrt{\dfrac{N-n}{N-1}} = \dfrac{25.53}{\sqrt{8}}\sqrt{\dfrac{50-8}{50-1}} = 8.36$. Thus the confidence interval is

$$\overline{x} - t_{0.005}\,\widehat{\sigma}_{\overline{x}} < \mu < \overline{x} + t_{0.005}\,\widehat{\sigma}_{\overline{x}}$$
$$188.63 - (3.499)(8.36) < \mu < 188.63 + (3.499)(8.36)$$
$$188.63 - 29.25 < \mu < 188.63 + 29.25$$
$$159.4 < \mu < 217.9.$$

7-4 Estimating the Population Percentage

2. $p = \dfrac{502}{862} \times (100 \text{ percent}) = 58.10$ percent and $\widehat{\sigma}_p = \sqrt{\dfrac{58.10 \cdot (100 - 58.10)}{864}} = 1.679$. The confidence interval is $58.10 \pm 1.645 \cdot 1.679$ or 55.34 to 60.86 percent.

4. $p = \dfrac{27}{46} \times (100 \text{ percent}) = 58.70$ percent and $\widehat{\sigma}_p = \sqrt{\dfrac{58.70 \cdot 41.3}{46}} = 7.26$. The confidence interval is $58.70 \pm 1.96 \cdot 7.26$ or 44.47 to 72.93 percent.

6. $p = \dfrac{124}{4,145} \times (100 \text{ percent}) = 2.99$ percent and $\widehat{\sigma}_p = \sqrt{\dfrac{2.99 \cdot 97.01}{4,145}} = 0.2645$. The confidence interval is $2.99 \pm 1.645 \cdot 0.2645$ or 2.55 to 3.43 percent.

8. $p = \dfrac{40}{63} \times (100 \text{ percent}) = 63.49$ percent and $\widehat{\sigma}_p = \sqrt{\dfrac{63.49 \cdot 36.51}{63}} = 6.07$. The confidence interval is $63.49 \pm 1.645 \cdot 6.07$ or 53.50 to 73.48 percent.

10. $p = \frac{11+61}{15+96} \times (100 \text{ percent}) = \frac{72}{111} \times (100 \text{ percent}) = 64.86$ percent and $\widehat{\sigma}_p = \sqrt{\frac{64.86 \cdot 35.14}{111}} = 4.53.$
The confidence interval is $64.86 \pm 1.96 \cdot 4.53$ or 55.98 to 73.74 percent.

12. $p = \frac{85}{210} \times (100 \text{ percent}) = 40.48$ percent and $\widehat{\sigma}_p = \sqrt{\frac{40.48 \cdot 59.52}{210}} = 3.39$. The confidence interval is $40.48 \pm 1.645 \cdot 3.39$ or 34.90 to 46.05 percent.

14. $p = \frac{115}{2142} \times (100 \text{ percent}) = 5.37$ percent and $\widehat{\sigma}_p = \sqrt{\frac{5.37 \cdot 94.63}{2142}} = 0.49$. The confidence interval is $5.37 \pm 1.645 \cdot 0.49$ or 4.56 to 6.18 percent.

16. $p = \frac{20}{163} \times (100 \text{ percent}) = 12.27$ percent and $\widehat{\sigma}_p = \sqrt{\frac{12.27 \cdot 87.73}{163}} = 2.57$. The confidence interval is $12.27 \pm 1.96 \cdot 2.57$ or 7.23 to 17.31 percent.

18. $p = \frac{24}{50} \times (100 \text{ percent}) = 48$ percent and $\widehat{\sigma}_p = \sqrt{\frac{48 \cdot 52}{50}} = 7.07$. The confidence interval is $48 \pm 1.645 \cdot 7.07$ or 36.37 to 59.63 percent.

20. $p = \frac{27}{51} \times (100 \text{ percent}) = 52.94$ percent and $\widehat{\sigma}_p = \sqrt{\frac{52.94 \cdot 47.06}{51}} = 6.99$. The confidence interval is $52.94 \pm 1.96 \cdot 6.99$ or 39.23 to 66.64 percent.

22. $p = \frac{39}{199} \times (100 \text{ percent}) = 19.60$ percent and $\widehat{\sigma}_p = \sqrt{\frac{19.6 \cdot 80.4}{199}} = 2.81$. The confidence interval is $19.60 \pm 2.575 \cdot 2.81$ or 12.36 to 26.84 percent.

7-5 Estimating the Population Variance

2. The lower and upper values are 39.4 and 12.40.

4. Using the χ^2 distribution with 15 df, the limits for the confidence interval of the variance are $\frac{15(26.8)}{32.8}$ and $\frac{15(26.8)}{4.60}$ or 12.256 to 87.39. Taking the square root of each variance, we find the confidence interval for the standard deviation is 3.50 to 9.35.

6. With 24 df, the limits for the confidence interval of the variance are $\frac{24(17.4)}{39.4}$ and $\frac{24(17.4)}{12.40}$ or 10.60 to 33.68. Taking the square root of each variance, we find the confidence interval for the standard deviation is 3.26 to 5.80. We are assuming that the population values are normally distributed.

8. With 26 df, the limits for the confidence interval of the variance are $\frac{26(14^2)}{48.3}$ and $\frac{26(14^2)}{11.16}$ or 105.51 to 456.63. Taking the square root of each variance, we find the confidence interval for the standard deviation is 10.27 to 21.37. Again, we are assuming that the population values are normally distributed.

10. Using MINITAB we find the standard deviation of the sample of 19 students is 51.7. With 18 df, the limits for the confidence interval of the variance are $\frac{18(51.7^2)}{37.2}$ and $\frac{18(51.7^2)}{6.26}$ or $1,293.33$ to $7,685.63$.

7-6 Determining Sample Size to Estimate μ or π

2. To be within 2 hours of the true population mean, we would need $n = \left[\frac{(1.96)(68.2)}{2}\right]^2 = 4{,}468$. That is, at least 4,468 in the sample. This means at least 4,468 − 100 or 4,368 more sample items.

4. At the 90 percent level, $z = 1.645$ and $\sigma = 4.346$. So our sample size must be at least $n = \left[\frac{(1.645)(4.346)}{0.5}\right]^2 = 204.44$ or 205 couples.

6. If all other factors remain the same but the confidence level is increased, the sample size must increase as well.

8. We can use the preliminary study and estimate π with $p = \frac{108}{160} = 67.5$ percent. So $n = \frac{(1.96)^2(67.5)(32.5)}{5^2} = 337.1$ or at least 338. Since the preliminary study had 160 syringes, we would need 338 − 160 or 178 more syringes.

10. Using $p = \frac{43}{319} = 13.48$ percent as an estimate for π, $n = \frac{(1.96)^2(13.48)(86.52)}{3^2} = 497.82$. The minimum sample size is 498. And we would need $498 - 319 = 179$ more male Vietnam veterans in our sample.

12. The minimum value for $n = \left[\frac{(1.645)(42.67)}{10}\right]^2 = 49.27$, or 50.

14. The minimum value for $n = \left[\frac{(2.575)(17)}{2}\right]^2 = 479.06$ or 480. Since there were 58 patients in the preliminary test, we would need 480 − 58 or 422 more for the sample.

16. At the 95 percent level, $z = 1.96$ and $\sigma = 11.675$. So our sample size must be at least $n = \left[\frac{(1.96)(11.675)}{2}\right]^2 = 130.9$ or 131 salmon.

Answers to Even-Numbered Exercises

2. Using the techniques of Chapter 6, 85 percent corresponds to a z-value of 1.44.

4. Using the techniques of Chapter 6, 98 percent corresponds to a z-value of 2.33.

6. With a 99 percent confidence level $\alpha = 0.01$ (1 percent). With a sample size of 6 there are 5 degrees of freedom. Therefore using Appendix 4 under $t_{0.005}$ we see t value is 4.032.

8. With a 95 percent confidence level $\alpha = 0.05$ (5 percent). With a sample size of 4 there are 3 degrees of freedom. Therefore using Appendix 4 under $t_{0.025}$ we see t value is 3.182.

10. We have the following data: $n = 21, \overline{x} = 23.1, s = 7.5$, population distribution shape is normal, $\sigma = 7.9$. Because the population distribution shape is normal and we know the value of s, we use the z distribution. 95 percent confidence level corresponds to $z = \pm 1.96$.

12. We have the following data: $n = 16, \overline{x} = 733.2, s = 45.5$, population distribution shape is normal, $\sigma = 41.7$. Because the population distribution shape is normal and we know the value of σ, we use the z distribution. 90 percent confidence level corresponds to $z = \pm 1.645$.

14. We have the following data: $n = 37, \overline{x} = 8.1, s = 3.9$, population distribution shape is unknown, σ unknown. Since $n > 30$, we use the z distribution and $z = \pm 1.645$.

16. Here $\alpha = 0.05$ so $\chi^2_{\alpha/2} = \chi^2_{0.025}$ and $\chi^2_{1-\alpha/2} = \chi^2_{0.975}$. Using Appendix 6 with $n - 1 = 12$ degrees of freedom we have $\chi^2_{0.975} = 4.40$ and $\chi^2_{0.025} = 23.3$.

18. We have the following data: $n = 49, \overline{x} = 76.1, s = 14.2$. Because $n > 30$, we use the z distribution, here $z = 1.96$. Thus $\widehat{\sigma}_{\overline{x}} = \dfrac{s}{\sqrt{n}} = \dfrac{14.2}{\sqrt{49}} = 2.029$. Hence the interval is $76.1 \pm 1.96(2.029)$ or 72.123 to 80.08.

20. The following table shows the results.

n:	36	49	64
Interval:	71.46 to 80.739	72.123 to 80.08	72.621 to 79.58

As the sample size increases the confidence interval width decreases.

22. We have the following data: $n = 100, \overline{x} = 364.1, s = 61.7$. Because $n > 30$, we use the z distribution, here $z = 1.96$. Thus $\widehat{\sigma}_{\overline{x}} = \dfrac{s}{\sqrt{n}} = \dfrac{61.7}{\sqrt{100}} = 6.17$. Hence the interval is $364.1 \pm 1.96(6.17)$ or 352.01 to 376.19.

24. The following table shows the results.

Percent:	90	95	99
Interval:	353.95 to 374.25	352.01 to 376.19	348.21 to 379.99

As level of confidence (percent) increases the confidence interval width also increases.

26. Substitute into the formula $\dfrac{(n-1)s^2}{\chi^2_{\alpha/2}} < \sigma^2 < \dfrac{(n-1)s^2}{\chi^2_{1-\alpha/2}}$. Data: $n = 24$ and $s^2 = 2.9$. Since $\alpha = 0.01$ with 23 df we have $\chi^2_{\alpha/2} = \chi^2_{0.005} = 44.2$ and $\chi^2_{1-\alpha/2} = \chi^2_{0.995} = 9.26$. Substituting

$$\frac{(n-1)s^2}{\chi^2_{\alpha/2}} < \sigma^2 < \frac{(n-1)s^2}{\chi^2_{1-\alpha/2}}$$

$$\frac{23 \cdot 2.9}{44.2} < \sigma^2 < \frac{23 \cdot 2.9}{9.26}$$

$$1.509 < \sigma^2 < 7.203.$$

28. $\widehat{\sigma}_p = \sqrt{\dfrac{63.79 \cdot 36.21}{116}} = 4.462.$

30. Point estimate $= \dfrac{\text{observed}}{\text{sample size}} = \dfrac{319}{1,190} = 26.81$ percent.

32. $z = \pm 1.96$ for a 95 percent confidence interval and $\widehat{\sigma}_p = \sqrt{\dfrac{26.81 \cdot 73.19}{1,190}} = 1.284$ percent, the interval is $26.81 \pm 1.96(1.284)$ or 24.29 to 29.33 percent.

34. To calculate the estimated standard error we construct the following table to the standard deviation of the sample using $\overline{x} = 44.71$ from Exercise 33.

x	$x-\overline{x}$	$(x-\overline{x})^2$	x	$x-\overline{x}$	$(x-\overline{x})^2$	x	$x-\overline{x}$	$(x-\overline{x})^2$
27	−17.71	313.6441	44	−0.71	0.5041	24	−20.71	428.9041
35	−9.71	94.2841	34	−10.71	114.7041	48	3.29	10.8241
65	20.29	411.6841	51	6.29	39.5641	29	−15.71	246.8041
67	22.29	496.8441	17	−27.71	767.8441	73	28.29	800.3241
47	2.29	5.2441	40	−4.71	22.1841	60	15.29	233.7841
46	1.29	1.6641	41	−3.71	13.7641	41	−3.71	13.7641
63	18.29	334.5241	60	15.29	233.7841	27	−17.71	313.6441
		1657.8887			1192.3487			2048.0487
							Total =	4898.2861

$s = \sqrt{\dfrac{\Sigma(x-\overline{x})^2}{n-1}} = \sqrt{\dfrac{4898.2861}{20}} = 15.65$. The estimated standard error, $\widehat{\sigma}_{\overline{x}} = \dfrac{s}{\sqrt{n}} = \dfrac{15.65}{\sqrt{21}} = 3.42$.

36. Data: $n = 21$ and $s^2 = 15.65^2 = 244.92$. $\alpha = 0.05$ and with 20 df we have $\chi^2_{\alpha/2} = \chi^2_{0.05} = 31.4$ and $\chi^2_{1-\alpha/2} = \chi^2_{0.95} = 10.85$. Substituting

$$\frac{(n-1)s^2}{\chi^2_{\alpha/2}} < \sigma^2 < \frac{(n-1)s^2}{\chi^2_{1-\alpha/2}}$$
$$\frac{20 \cdot 244.92}{31.4} < \sigma^2 < \frac{20 \cdot 244.92}{10.85}$$
$$156 < \sigma^2 < 451.47.$$

38. To calculate the estimated standard error we construct the following table to the standard deviation of the sample using $\overline{x} = 13.24$ from Exercise 37.

x	$x-\overline{x}$	$(x-\overline{x})^2$	x	$x-\overline{x}$	$(x-\overline{x})^2$	x	$x-\overline{x}$	$(x-\overline{x})^2$
9.7	−3.54	12.5316	12.8	−0.44	0.1936	19.1	5.86	34.3396
21.0	7.76	60.2176	8.6	−4.64	21.5296	7.0	−6.24	38.9376
14.3	1.06	1.1236	10.9	−2.34	5.4756	19.5	6.26	39.1876
15.2	1.96	3.8416	8.3	−4.94	24.4036	12.5	−0.74	0.5476
		77.7144			51.6024			113.0124
							Total =	242.3292

$s = \sqrt{\dfrac{\Sigma(x-\overline{x})^2}{n-1}} = \sqrt{\dfrac{242.3292}{11}} = 4.69$. The estimated standard error, $\widehat{\sigma}_{\overline{x}} = \dfrac{s}{\sqrt{n}} = \dfrac{4.69}{\sqrt{12}} = 1.35$.

40. Data: $n = 12$ and $s^2 = 22.00$. $\alpha = 0.01$ and with 11 df we have $\chi^2_{\alpha/2} = \chi^2_{0.005} = 26.8$ and $\chi^2_{1-\alpha/2} = \chi^2_{0.995} = 2.60$. Substituting

$$\frac{(n-1)s^2}{\chi^2_{\alpha/2}} < \sigma^2 < \frac{(n-1)s^2}{\chi^2_{1-\alpha/2}}$$
$$\frac{11 \cdot 22.00}{26.8} < \sigma^2 < \frac{11 \cdot 22.00}{2.60}$$
$$9.030 < \sigma^2 < 93.08$$
$$3.005 < \sigma < 9.65.$$

42. To calculate the estimated standard error we construct the following table to the standard deviation of the sample using $\overline{x} = \$2,996.67$ from Exercise 41.

x	$x-\overline{x}$	$(x-\overline{x})^2$	x	$x-\overline{x}$	$(x-\overline{x})^2$	x	$x-\overline{x}$	$(x-\overline{x})^2$
$3,259	262.33	68817.03	3,295	298.33	89000.79	2,882	−114.67	13149.21
3,133	136.33	18585.87	3,025	28.33	802.59	2,448	−548.67	301038.77
3,465	468.33	219332.99	2,673	−323.67	104762.27	2,939	−57.67	3325.83
2,963	−33.67	1133.67	3,254	257.33	66218.73	2,624	−372.67	138882.93
		307869.56			260784.38			456396.74
							Total =	1025050.68

$s = \sqrt{\dfrac{\Sigma(x-\overline{x})^2}{n-1}} = \sqrt{\dfrac{1025050.68}{11}} = \$305.265$. The estimated standard error, $\widehat{\sigma}_{\overline{x}} = \dfrac{s}{\sqrt{n}} = \dfrac{305.265}{\sqrt{12}} = \$88.122$.

44. Data: $n = 12$ and $s^2 = 93186.72$. $\alpha = 0.05$ and with 11 df we have $\chi^2_{\alpha/2} = \chi^2_{0.025} = 21.9$ and $\chi^2_{1-\alpha/2} = \chi^2_{0.975} = 3.82$. Substituting

$$\frac{(n-1)s^2}{\chi^2_{\alpha/2}} < \sigma^2 < \frac{(n-1)s^2}{\chi^2_{1-\alpha/2}}$$

$$\frac{11 \cdot 93186.72}{21.9} < \sigma^2 < \frac{11 \cdot 93186.72}{3.82}$$

$$46{,}806.12 < \sigma^2 < 268{,}338.72.$$

46. We have the following data: $n = 26$, $\overline{x} = 97.6$, population distribution shape normal, $\sigma = 15$. Since the population shape is normal and σ is known we use the z distribution. $z = \pm 1.645$ and $\sigma_{\overline{x}} = \dfrac{\sigma}{\sqrt{n}} = \dfrac{15}{\sqrt{26}} = 2.94$ Thus $\overline{x} \pm z \cdot \sigma_{\overline{x}} = 97.6 \pm 1.645(2.94)$ or an IQ of 92.76 to 102.44.

48. Here we want to find n so that $z\widehat{\sigma}_p < 2$ percent when $p = \dfrac{99}{431} \times (100 \text{ percent}) = 22.97$ percent, $z = 1.645$, and $n = \dfrac{(1.645)^2(22.97)(77.03)}{2^2} = 1,196.99$. Thus the sample size must be at least 1197.

50. Since the sample size is less than 30 and the population shape is normal, we use the t distribution. We have the following data: $n = 10, \overline{x} = 147.8$, $s = 31.2$. Then $t_{0.025}$ with 9 df is 2.262 and $\widehat{\sigma}_{\overline{x}} = \dfrac{s}{\sqrt{n}} = \dfrac{31.2}{\sqrt{10}} = 9.87$. The interval is $147.8 \pm 2.262(9.87)$ or 125.47 to 170.13.

52. Data: $n = 36, \overline{x} = \dfrac{24}{36} \times (100 \text{ percent}) = 66.67$ percent. So $z = \pm 1.645$ and $\widehat{\sigma}_p = \sqrt{\dfrac{66.67 \cdot 33.33}{36}} = 7.86$ percent. The interval is $66.67 \pm 1.645(7.86)$ or 53.74 to 79.60 percent.

54. Sample size > 30, use the z distribution. Data: $n = 281, \overline{x} = 58.3$, s = 20.5. So $z = \pm 1.96$ and $\widehat{\sigma}_{\overline{x}} = \dfrac{s}{\sqrt{n}} = \dfrac{20}{\sqrt{281}} = 1.223$. The interval is $58.3 \pm 1.96(1.223)$ or 55.90 to 60.70 words.

56. We are given the following data: $n = 64, \overline{x} = 24.70, s = 5.14$. Because $n > 30$ we use the z distribution, here $z = \pm 2.575$. Thus $\widehat{\sigma}_{\overline{x}} = \dfrac{s}{\sqrt{n}} = \dfrac{5.14}{\sqrt{64}} = 0.64$. Hence the interval is $24.70 \pm 2.575(0.64)$ or 23.05 to 26.35.

58. Find n so that $z\widehat{\sigma}_{\overline{x}} < 1$ kg when $z = 1.96$ and $\widehat{\sigma}_{\overline{x}} = \dfrac{s}{\sqrt{n}} = \dfrac{3.6}{\sqrt{n}}$. Then $n = \left[\dfrac{(1.96)(3.6)}{1}\right]^2 = 49.79$. Thus the sample size must contain at least 50 participants.

60. Find n so that $z\widehat{\sigma}_{\overline{x}} < 2$ hours when $s = 15.2$ hours. Using $z = 1.645$ and $\widehat{\sigma}_{\overline{x}} = \frac{s}{\sqrt{n}} = \frac{15.2}{\sqrt{n}}$, then

$n = \left[\frac{(1.645)(15.2)}{2}\right]^2 = 156.3$. Thus the sample size must contain at least 157 calculator batteries.

62. Find n so that $z\,\widehat{\sigma}_p < 5$ percent when π is unknown, $z = 2.575$, and $\widehat{\sigma}_p$ is maximized when $p = 1 - p$, use 50 percent as our point estimate. Then $n = \frac{(2.575)^2(50)(50)}{5^2} = 663.1$. Thus the sample size must contain at least 664 products.

64. Using $\overline{x} = 1593.7$ and $s = 235.83$ from Exercise 63, since df $= 11$, $t_{0.005} = \pm 3.106$. Then $\widehat{\sigma}_{\overline{x}} = \frac{s}{\sqrt{n}} = \frac{235.83}{\sqrt{12}} = 68.08$. Thus the confidence interval for μ is

$$\overline{x} - t_{0.005}\,\widehat{\sigma}_{\overline{x}} < \mu < \overline{x} + t_{0.005}\,\widehat{\sigma}_{\overline{x}}$$
$$1593.7 - (3.106)(68.08) < \mu < 1593.7 + (3.106)(68.08)$$
$$1593.7 - 211.46 < \mu < 1593.7 + 211.46$$
$$1382.24 < \mu < 1805.16.$$

66. Data: $n = 12$ and $s^2 = (235.83)^2 = 55615.79$. $\alpha = 0.01$ and with 11 df we have $\chi^2_{\alpha/2} = \chi^2_{0.005} = 26.8$ and $\chi^2_{1-\alpha/2} = \chi^2_{0.995} = 2.60$. Substituting

$$\frac{(n-1)s^2}{\chi^2_{\alpha/2}} < \sigma^2 < \frac{(n-1)s^2}{\chi^2_{1-\alpha/2}}$$
$$\frac{11 \cdot 55615.79}{26.8} < \sigma^2 < \frac{11 \cdot 55615.79}{2.60}$$
$$22{,}827.38 < \sigma^2 < 235{,}297.57$$
$$151.09 < \sigma < 485.07.$$

Chapter 8 Testing Hypotheses: One-Sample Procedures

Answers to Even-Numbered Self-Testing Review

8-1 The Hypothesis-Testing Procedure in General

2. The sampling distribution in this case is a t distribution with 13 df. A sample mean with $n = 14$ has a 0.02 chance of falling beyond ± 2.65 standard errors.

4. You'll make a Type I error when the null hypothesis is true but you decide to reject it.

6. It is the probability of making a Type I error (rejecting a true null hypothesis).

8. You need to know the correct distribution to determine (from the appropriate table) the critical value that starts the rejection region.

10. The test statistic is the difference between the parameter value in the null hypothesis and the corresponding sample statistic value converted to a standard score.

12. The hypothesis that you accept when the H_0 is rejected.

14. When you test the hypothesis that the mean is different than 100, you don't know if it's more or less, so this must be $\mu \neq 100$. Now we need a hypothesis that contains the $=$ symbol, so this is $\mu = 100$. We choose the null hypothesis as the one with the $=$ symbol, so this will be a two-tailed test with: H_0: $\mu = 100$, and H_1: $\mu \neq 100$.

16. The hypothesis that sales have increased is $\mu >$ \$85,492, and a statement of equality is needed for the null hypothesis. Thus, H_0: $\mu =$ \$85,492, and H_1: $\mu >$ \$85,492.

18. The hypothesis of interest is $\mu < 14.09$. Since it doesn't contain a statement of equality, it must be the alternative hypothesis. Thus, H_0: $\mu = 14.09$, and H_1: $\mu < 14.09$.

8-2 One-Sample Hypothesis Tests of Means

2. For a two-tailed test, subtract $1 - 0.05$ to get 0.95 which is the area from $-z$ to $+z$. Now take half of that value to read the corresponding area values. That is, $\frac{0.95}{2} = 0.4750$. And this corresponds to a z score of 1.96. For a two-tailed test, we use ± 1.96.

4. For $n - 1 = 12$ degrees of freedom, the corresponding left tailed value is $t = -1.782$.

6. For 8 df under the column headed by 0.05, we see that $t = 1.860$.

8. Use the value of $z = 2.33$.

10. Use the t distribution with 11 df, so $t = -1.796$.

12. The decision rule is to reject the H_0 when the p-value is less than the level of significance, and fail to reject the H_0 when the p-value is $\geq \alpha$.

14. The p-value is $< \alpha$, so you reject the H_0.

16. Since 0.002 is < 0.01 and 0.05, the decision is to reject H_0 at both levels.

18. Step 1: State the null and alternative hypotheses. H_0: $\mu = 50$, and H_1: $\mu \neq 50$.
Step 2: Select the level of significance. $\alpha = 0.05$.
Step 3: Determine the test distribution to use. The z distribution is used for a large sample.
Step 4: Define the critical or rejection region(s). The values for a two-tailed test at the 0.05 level are $z = -1.96$ and $z = +1.96$.
Step 5: State the decision rule. Reject H_0 in favor of H_1 if $z < -1.96$ or $z > +1.96$. Otherwise, fail to reject H_0.
Step 6: Compute the Test Statistic. The standard error is $\frac{5.1}{\sqrt{45}} = 0.7603$, so the
$z = \frac{48.3 - 50}{0.7603} = -2.24$.
Step 7: Make the statistical decision. Since the z value of -2.24 falls in the left tailed rejection region, we reject H_0.

20. Step 1: State the null and alternative hypotheses. H_0: $\mu = 1{,}500$, and H_1: $\mu < 1{,}500$.
Step 2: Select the level of significance. $\alpha = 0.05$
Step 3: Determine the test distribution to use. The z distribution is used.
Step 4: Define the critical or rejection region(s). The value for this one-tailed test at the 0.05 level is $z = -1.645$.
Step 5: State the decision rule. Reject H_0 in favor of H_1 if $z < -1.645$. Otherwise, fail to reject H_0.
Step 6: Compute the Test Statistic. The standard error is $\frac{51.3}{\sqrt{62}} = 6.515$, so $z = \frac{1{,}483 - 1{,}500}{6.515} = -2.61$.
Step 7: Make the statistical decision. Since -2.61 falls in the left tailed rejection region, we reject H_0.

22. Step 1: H_0: $\mu = \$5{,}423$, and H_1: $\mu > \$5{,}423$.
Step 2: $\alpha = 0.01$.
Step 3: The z distribution is used.
Step 4: The value for a right-tailed test at the 0.01 level is $z = 2.33$.
Step 5: Reject H_0 in favor of H_1 if $z > 2.33$. Otherwise, fail to reject H_0.
Step 6: The $z = \frac{\$5{,}516 - \$5{,}423}{\$52.1809} = 1.78$.
Step 7: Since the z value of 1.78 does not fall in the rejection region, we fail to reject H_0, the average expense of Seattle residents over 64 is \$5,423.

24. Step 1: H_0: $\mu = 1.30$, and H_1: $\mu > 1.30$.
Step 2: $\alpha = 0.05$.
Step 3: The z distribution is used.
Step 4: The value for a right-tailed test at the 0.05 level is $z = 1.645$.

Step 5: Reject H_0 in favor of H_1 if $z > 1.645$. Otherwise, fail to reject H_0.
Step 6: The $z = \dfrac{1.87 - 1.30}{0.2468} = 2.31$.
Step 7: Since the z value of 2.31 falls in the rejection region, we reject H_0 and decide at $\alpha = 0.05$ that Project CALC students do better.

26. Step 1: H_0: $\mu = \$31{,}129$, and H_1: $\mu \neq \$31{,}129$.
Step 2: $\alpha = 0.05$.
Step 3: Use the t distribution with 14 df.
Step 4: The values for a two-tailed test at the 0.05 level are $t = \pm 2.145$.
Step 5: Reject H_0 in favor of H_1 if $t < -2.145$ or $t > +2.145$. Otherwise, fail to reject H_0.
Step 6: The $t = \dfrac{\$32{,}379 - \$31{,}129}{\$463.98} = 2.69$.
Step 7: Since the t value of 2.69 falls in the rejection region, we reject H_0, and conclude that the mean for the Denver firms differs from \$31,129.

28. Step 1: H_0: $\mu = 84.2$, and H_1: $\mu < 84.2$.
Step 2: $\alpha = 0.01$.
Step 3: Use the t distribution with 28 df.
Step 4: The values for a left-tailed test at the 0.01 level $t = -2.467$.
Step 5: Reject H_0 in favor of H_1 if $t < -2.467$. Otherwise, fail to reject H_0.
Step 6: The $t = \dfrac{69.1 - 84.2}{2.90} = -5.21$.
Step 7: Since the t value of -5.21 falls in the rejection region, we reject H_0 and decide that the patients who take *Captopril* have lower blood pressure than those who do not.

30. Step 1: H_0: $\mu = 1{,}400$, and H_1: $\mu \neq 1{,}400$.
Step 2: $\alpha = 0.05$.
Step 3: Use the t distribution with 10 df.
Step 4: Reject H_0 in favor of H_1 if the p-value is < 0.05. Otherwise, fail to reject H_0.
Step 5: The $t = \dfrac{1{,}381.8 - 1{,}400}{67.2} = -0.27$.
Step 6: With 10 df, a t value of $+2.228$ has a p value of 0.05. Since -0.27 is closer to $t = 0$, it has a p value that is greater than 0.05.
Step 7: Since the p-value for this test is > 0.05, we fail to reject H_0 and decide that the mean impedance is 1,400.

32. Step 1: H_0: $\mu = 3.9$, and H_1: $\mu < 3.9$.
Step 2: $\alpha = 0.01$.
Step 3: Use the z distribution.
Step 4: Reject H_0 in favor of H_1 if the p-value is < 0.01. Otherwise, fail to reject H_0.
Step 5: The $z = \dfrac{3.78 - 3.9}{0.1123} = -1.07$.
Step 6: The area to the left of $z = -1.07$ is $0.5 - 0.3577$ or 0.1423. Thus, $p = 0.1423$.
Step 7: The p-value for this test is greater than 0.01. Following the decision rule, we fail to reject H_0.

34. Step 1: H_0: $\mu = 350$, and H_1: $\mu > 350$.
Step 2: $\alpha = 0.05$.
Step 3: Use the t distribution with 9 df.
Step 4: The values for a right-tailed test at the 0.05 level, $t = 1.833$.
Step 5: Reject H_0 in favor of H_1 if $t > 1.833$. Otherwise, fail to reject H_0.

Step 6: Calculate $\overline{x}$ and s.

x	$x-\overline{x}$	$(x-\overline{x})^2$	x	$x-\overline{x}$	$(x-\overline{x})^2$
401	−6.3	39.69	415	7.7	59.29
359	−48.3	2332.89	389	−18.3	334.89
383	−24.3	590.49	463	55.7	3102.49
427	19.7	388.09	394	−13.3	176.89
414	6.7	44.89	428	20.7	428.49
1984		3396.05	2089		4102.05

$\Sigma x = 1984 + 2089 = 4073$. $\Sigma(x-\overline{x})^2 = 3396.05 + 4102.05 = 7498.10$

$\overline{x} = \dfrac{4073}{10} = 407.3$. $s = \sqrt{\dfrac{7498.10}{9}} = 28.86$. Since σ is unknown we use $\widehat{\sigma}_{\overline{x}} = \dfrac{s}{\sqrt{n}} = \dfrac{28.86}{\sqrt{10}} = 9.13$.

Then $t = \dfrac{407.3 - 350}{9.13} = 6.28$.

Step 7: Since the t value of 6.28 falls in the rejection region, we reject H_0 and decide that the mean bursting pressure of more than 350 psi.

36. Step 1: H_0: $\mu = 38,000$, and H_1: $\mu > 38,000$.
Step 2: $\alpha = 0.10$.
Step 3: Use the z distribution.
Step 4: The values for a right-tailed test at the 0.10 level, $z = 1.28$.
Step 5: Reject H_0 in favor of H_1 if $z > 1.28$. Otherwise, fail to reject H_0.
Step 6: From MINITAB, $\overline{x} = 39,896$ and $s = 8959$. Since σ is unknown we use
$\widehat{\sigma}_{\overline{x}} = \dfrac{s}{\sqrt{n}} = \dfrac{8959}{\sqrt{31}} = 1609.08$. Then $z = \dfrac{39,896 - 38,000}{1609.08} = 1.18$.
Step 7: Since the z value of 1.18 does not fall in the rejection region, we fail to reject H_0, there is not significant evidence that the mean number of hits on all days exceeds 38,000/day.

38. Step 1: H_0: $\mu = 0.33$, and H_1: $\mu \neq 0.33$.
Step 2: $\alpha = 0.05$.
Step 3: Use the z distribution.
Step 4: The values for a right-tailed test at the 0.05 level, $z = \pm 1.96$.
Step 5: Reject H_0 in favor of H_1 if $z < -1.96$ or if $z > 1.96$. Otherwise, fail to reject H_0.
Step 6: From MINITAB, $\overline{x} = 0.33127$ and $s = 0.01207$. Since σ is unknown we use
$\widehat{\sigma}_{\overline{x}} = \dfrac{s}{\sqrt{n}} = \dfrac{0.01207}{\sqrt{30}} = 0.00220$. Then $z = \dfrac{0.33127 - 0.33}{0.00220} = 0.577$.
Step 7: Since the z value of 0.577 does not fall in the rejection region, we fail to reject H_0, there is not significant evidence that the nominal value of the capacitors is 0.33 μF.

8-3 One-Sample Hypotheses Tests of Percentages

2. Step 1: The statement in the problem translates to the hypothesis that $\pi = 5$ percent which must be the null hypothesis since it contains the statement of equality. The alternative must be the two-tailed version since no direction is indicated. Thus, H_0: $\pi = 5$ percent, and H_1: $\pi \neq 5$ percent.
Step 2: $\alpha = 0.05$.
Step 3: Use the z distribution.
Step 4: The values for a two-tailed test at the 0.05 level are $z = \pm 1.96$.
Step 5: Reject H_0 in favor of H_1 if $z < -1.96$ or $z > +1.96$. Otherwise, fail to reject H_0.
Step 6: The $z = \dfrac{13.0952 - 5}{2.3780} = 3.40$.

Step 7: Since the z value of 3.40 falls in the rejection region, we reject H_0.

4. Step 1: H_0: $\pi = 75$ percent, and H_1: $\pi > 75$ percent.
Step 2: $\alpha = 0.01$.
Step 3: Use the z distribution.
Step 4: The value for a right-tailed test at the 0.05 level is $z = 2.33$.
Step 5: Reject H_0 in favor of H_1 if $z > 2.33$. Otherwise, fail to reject H_0.
Step 6: $z = \dfrac{77.0155 - 75}{4.6424} = 0.4333$.
Step 7: Since the z value of 0.4333 does not fall in the rejection region, we fail to reject H_0 that the population percentage is 75 percent.

6. Step 1: H_0: $\pi = 15$ percent, and H_1: $\pi \neq 15$ percent.
Step 2: $\alpha = 0.01$.
Step 3: Use the z distribution.
Step 4: The value for a two-tailed test at the 0.01 level is $z = \pm 2.575$.
Step 5: Reject H_0 in favor of H_1 if $z < -2.575$ or $z > +2.575$. Otherwise, fail to reject H_0.
Step 6: $z = \dfrac{14.0502 - 15}{0.9560} = -0.9935$.
Step 7: Since the z of -0.9935 does not fall in the rejection region, we fail to reject H_0.

8. Step 1: H_0: $\pi = 28.4$ percent, and H_1: $\pi < 28.4$ percent.
Step 2: $\alpha = 0.01$.
Step 3: Use the z distribution.
Step 4: The value for a left-tailed test at the 0.01 level is $z = -2.33$.
Step 5: Reject H_0 in favor of H_1 if $z < -2.33$. Otherwise, fail to reject H_0.
Step 6: $z = \dfrac{10.0890 - 28.4}{2.4564} = -7.4544$.
Step 7: Since the z value of -7.4544 falls far out in the rejection region, we reject H_0.

10. Step 1: H_0: $\pi = 50$ percent, and H_1: $\pi > 50$ percent.
Step 2: $\alpha = 0.01$.
Step 3: Use the z distribution.
Step 4: The value for a right-tailed test at the 0.10 level is $z = 1.28$.
Step 5: Reject H_0 in favor of H_1 if $z > 1.28$. Otherwise, fail to reject H_0.
Step 6: $p = \dfrac{30}{47} = 63.83$ percent and $\widehat{\sigma}_p = \sqrt{\dfrac{50 \cdot 50}{47}} = 7.29$. $z = \dfrac{63.83 - 50}{7.29} = 1.90$.
Step 7: Since the z value of 1.90 falls in the rejection region, we reject H_0.

12. Step 1: H_0: $\pi = 20$ percent, and H_1: $\pi \neq 20$ percent.
Step 2: $\alpha = 0.05$.
Step 3: Use the z distribution.
Step 4: The value for a two-tailed test at the 0.05 level is $z = \pm 1.96$.
Step 5: Reject H_0 in favor of H_1 if $z < -1.96$ or $z > +1.96$. Otherwise, fail to reject H_0.
Step 6: $p = \dfrac{22}{143} = 15.38$ percent and $\widehat{\sigma}_p = \sqrt{\dfrac{20 \cdot 80}{143}} = 3.34$. $z = \dfrac{15.38 - 20}{3.34} = -1.38$.
Step 7: Since the z of -1.38 does not fall in the rejection region, we fail to reject H_0.

14. Step 1: H_0: $\pi = 20$, and H_1: $\pi > 20$.
Step 2: $\alpha = 0.10$.

Step 3: Use the z distribution.
Step 4: Reject H_0 in favor of H_1 if the p-value is < 0.10. Otherwise, fail to reject H_0.
Step 5: $p = 38.46$ percent and $\sigma_p = 7.845$. Then $z = \dfrac{38.46 - 20}{7.845} = 2.35$.
Step 6: The area to the right of $z = 2.35$ is $0.5 - 0.4906$ or 0.0094. Thus, p-value $= 0.0094$.
Step 7: The p-value for this test is less than 0.10. Following the decision rule, we reject H_0.

16. Step 1: H_0: $\pi = 80$ percent, and H_1: $\pi < 80$ percent.
Step 2: $\alpha = 0.05$.
Step 3: Use the z distribution.
Step 4: The value for a left-tailed test at the 0.05 level is $z = 1.645$.
Step 5: Reject H_0 in favor of H_1 if $z < -1.645$. Otherwise, fail to reject H_0.
Step 6: $p = \dfrac{67}{115} = 58.26$ percent and $\sigma_p = \sqrt{\dfrac{80 \cdot 20}{115}} = 3.73$. $z = \dfrac{58.26 - 80}{3.73} = -5.83$.
Step 7: Since the z of -5.83 falls in the rejection region, we reject H_0 in favor of H_1.

18. Step 1: H_0: $\pi = 80$ percent, and H_1: $\pi < 80$ percent.
Step 2: $\alpha = 0.05$.
Step 3: Use the z distribution.
Step 4: The value for a left-tailed test at the 0.05 level is $z = 1.645$.
Step 5: Reject H_0 in favor of H_1 if $z < -1.645$. Otherwise, fail to reject H_0.
Step 6: $p = \dfrac{83}{143} = 58.04$ percent and $\sigma_p = \sqrt{\dfrac{80 \cdot 20}{143}} = 3.34$. $z = \dfrac{58.04 - 80}{3.34} = -6.575$.
Step 7: Since the z of -6.575 falls in the rejection region, we reject H_0 in favor of H_1.

20. Step 1: H_0: $\pi = 28$, and H_1: $\pi \neq 28$.
Step 2: $\alpha = 0.05$.
Step 3: Use the z distribution.
Step 4: Reject H_0 in favor of H_1 if the p-value is < 0.025. Otherwise, fail to reject H_0.
Step 5: $p = \dfrac{82}{222} = 36.94$ percent and $\sigma_p = \sqrt{\dfrac{28 \cdot 72}{222}} = 3.013$. The $z = \dfrac{36.94 - 28}{3.012} = 2.97$.
Step 6: The area to the left of $z = 2.97$ is $0.5 - 0.4985 = 0.0015$. The p-value $= 2(0.0015) = 0.0030$.
Step 7: The p-value for this test is less than 0.05. Following the decision rule, we reject H_0 in favor of H_1.

8-4 One-Sample Hypothesis Tests of Variances and Standard Deviations

2. Use χ^2 with $7 - 1$ or 6 df. Since the critical region must contain a total of 0.01 in two tails, each tail must have 0.005. For the left critical value, look under the column headed 0.995 to find $\chi^2 = 0.676$, and for the right critical value look under the column headed 0.005 to find $\chi^2 = 18.55$.

4. Step 1: H_0: $\sigma^2 = 85$, and H_1: $\sigma^2 \neq 85$.
Step 2: $\alpha = 0.01$.
Step 3: Use the χ^2 distribution with 15 df.
Step 4: For a two-tailed test, there must be half of 0.01 or 0.005 in each tail. Looking under the 0.995 column for the left critical χ^2 value, and under 0.005 for the right value. These values are 4.60 and 32.8.
Step 5: Reject H_0 in favor of H_1 if $\chi^2 < 4.60$ or if $\chi^2 > 32.8$. Otherwise, fail to reject H_0.
Step 6: The $\chi^2 = \dfrac{15(78)}{85} = 13.765$.

Step 7: Since χ^2 of 13.765 does not fall in a rejection region, we fail to reject H_0.

6. Step 1: H_0: $\sigma = 4.2$, and H_1: $\sigma < 4.2$.
Step 2: $\alpha = 0.05$.
Step 3: Use the χ^2 distribution with 26 df.
Step 4: For a left-tailed test, look under the 0.95 column to find $\chi^2 = 15.38$.
Step 5: Reject H_0 in favor of H_1 if $\chi^2 < 15.38$. Otherwise, fail to reject H_0.
Step 6: The $\chi^2 = 26 \cdot \frac{3.9^2}{4.2^2} = 22.4184$.
Step 7: Since χ^2 of 22.4184 does not fall in the rejection region, we fail to reject H_0.

8. Step 1: H_0: $\sigma^2 = 0.5$, and H_1: $\sigma^2 < 0.5$.
Step 2: $\alpha = 0.01$.
Step 3: Use the χ^2 distribution with 15 df.
Step 4: $\chi^2 = 5.23$.
Step 5: Reject H_0 in favor of H_1 if $\chi^2 < 5.23$. Otherwise, fail to reject H_0.
Step 6: The $\chi^2 = \frac{15(0.35)}{0.5} = 10.5$.
Step 7: Since χ^2 of 10.5 does not fall in the rejection region, we fail to reject H_0.

10. Step 1: H_0: $\sigma = 40$, and H_1: $\sigma > 40$.
Step 2: $\alpha = 0.05$.
Step 3: Use the χ^2 distribution with 23 df.
Step 4: $\chi^2 = 35.2$.
Step 5: Reject H_0 in favor of H_1 if $\chi^2 > 35.2$. Otherwise, fail to reject H_0.
Step 6: The $\chi^2 = 11 \cdot \frac{113^2}{90^2} = 36.2577$.
Step 7: Since χ^2 of 36.2577 falls in the rejection region, we reject H_0.

12. Step 1: H_0: $\sigma^2 = 3.5$, and H_1: $\sigma^2 > 3.5$.
Step 2: $\alpha = 0.05$.
Step 3: Use the χ^2 distribution with 7 df.
Step 4: $\chi^2 = 14.07$.
Step 5: Reject H_0 in favor of H_1 if $\chi^2 > 14.07$. Otherwise, fail to reject H_0.
Step 6: The $\chi^2 = 7 \cdot \frac{2^2}{3.5} = 8$.
Step 7: Since χ^2 of 8 does not fall in the rejection region, we fail to reject H_0.

14. Step 1: H_0: $\sigma = 1$, and H_1: $\sigma > 1$.
Step 2: $\alpha = 0.01$.
Step 3: Use the χ^2 distribution with 11 df.
Step 4: $\chi^2 = 24.7$.
Step 5: Reject H_0 in favor of H_1 if $\chi^2 > 24.7$. Otherwise, fail to reject H_0.
Step 6: We calculate s^2.

x	x^2	x	x^2	x	x^2
−1.99	3.9601	−0.80	0.6400	0.27	0.0729
1.24	1.5376	2.13	4.5369	2.97	8.8209
−1.37	1.8769	−0.25	0.0625	0.76	0.5776
1.69	2.8561	2.56	6.5536	3.42	11.6964
−0.43	10.2307	3.64	11.7930	7.42	21.1678

$\Sigma x = -0.43 + 3.64 + 7.42 = 10.63$; $\Sigma x^2 = 10.2307 + 11.7930 + 21.1678 = 43.1915$. So

$s^2 = \dfrac{n(\Sigma x^2) - (\Sigma x)^2}{n(n-1)} = \dfrac{12(43.1915) - (10.63)^2}{12(11)} = 3.0704$. Then $\chi^2 = \dfrac{11(3.0704)}{1^2} = 33.77$.

Step 7: Since χ^2 of 33.77 falls in the rejection region, we reject H_0.

Answers to Even-Numbered Exercises

2. Step 1: State the Null and Alternative Hypotheses. H_0: $\mu = 58$, and H_1: $\mu \neq 58$.
Step 2: Select the Level of Significance. $\alpha = 0.05$ (given).
Step 3: Determine the Test Distribution to Use. Since $n < 30$ and we assume that the population values are normally distributed we use the t distribution.
Step 4: Define the Rejection or Critical Regions. This is a two-tailed test. With $\alpha = 0.05$ and $n = 24$ we have 23 df so we use $t_{\alpha/2} = t_{0.025} = \pm 2.069$.
Step 5: State the Decision Rule. Reject H_0 in favor of H_1 if $t > 2.069$ or if $t < -2.069$. Otherwise, fail to reject H_0.
Step 6: Make the Necessary Computations. Data: $\overline{x} = 55.93$; $s = 5.2$; $n = 24$. Since σ is unknown we use
$\hat{\sigma}_{\overline{x}} = \dfrac{s}{\sqrt{n}} = \dfrac{5.2}{\sqrt{24}} = 1.0614$. Then $t = \dfrac{\overline{x} - \mu_0}{\hat{\sigma}_{\overline{x}}} = \dfrac{55.93 - 58}{1.0614} = -1.950$.
Step 7: Make a Statistical Decision. Since $t = -1.950$ falls between ± 2.069 we fail to reject H_0.

4. Step 1: H_0: $\sigma = 85.2$, and H_1: $\sigma \neq 85.2$.
Step 2: $\alpha = 0.01$ (given).
Step 3: Use χ^2 test on variances.
Step 4: This is a two-tailed test, the acceptance region falls between $\chi^2_{\alpha/2}$ and $\chi^2_{1-\alpha/2}$. With $\alpha = 0.01$ and $n = 28$ we have 27 df so $\chi^2_{\alpha/2} = \chi^2_{0.005} = 49.6$, while $\chi^2_{1-\alpha/2} = \chi^2_{0.995} = 11.81$.
Step 5: Reject H_0 in favor of H_1 if $\chi^2 > 49.6$ or if $\chi^2 < 11.81$. Otherwise, fail to reject H_0.
Step 6: Data: $s = 81.7$; $n = 28$. So $s^2 = 6674.89$ and $\sigma^2 = 7259.04$.
$\chi^2 = \dfrac{(n-1)s^2}{\sigma^2} = \dfrac{27 \cdot 6674.89}{7259.04} = 24.827$.
Step 7: Since χ^2 falls between $\chi^2_{\alpha/2}$ and $\chi^2_{1-\alpha/2}$ we fail to reject H_0.

6. Step 1: H_0: $\sigma = \$150$, and H_1: $\sigma > \$150$.
Step 2: $\alpha = 0.05$.
Step 3: Use χ^2 test on variances.
Step 4: This is a one-tailed test, the acceptance region falls below χ^2_α. With $n = 19$ we have 18 df so $\chi^2_\alpha = \chi^2_{0.05} = 28.9$.
Step 5: Reject H_0 in favor of H_1 if $\chi^2 > 28.9$. Otherwise, fail to reject H_0.
Step 6: Calculate $\overline{x}$ and s^2.

x	$x-\overline{x}$	$(x-\overline{x})^2$	x	$x-\overline{x}$	$(x-\overline{x})^2$	x	$x-\overline{x}$	$(x-\overline{x})^2$
499	160.79	25853.4241	100	−238.21	56744.0041	489	150.79	22737.6241
279	−59.21	3505.8241	235	−103.21	10652.3041	300	−38.21	1460.0041
669	330.79	109422.0241	467	128.79	16586.8641	299	−39.21	1537.4241
550	211.79	44855.0041	249	−89.21	7958.4241	200	−138.21	19102.0041
207	−131.21	17216.0641	200	−138.21	19102.0041	200	−138.21	19102.0041
600	261.79	68534.0041	235	−103.21	10652.3041	249	−89.21	7958.4241
399	60.79	3695.4241						
3203		273081.7687	1486		121695.9046	1737		71897.4846

$\Sigma x = 3203 + 1486 + 1737 = 6426$, $\Sigma x^2 = 273081.7687 + 121695.9046 + 71897.4846 = 466675.1579$,

$\overline{x} = \dfrac{6426}{19} = \$338.21$, $s^2 = \sqrt{\dfrac{\Sigma(x-\overline{x})^2}{n-1}} = \sqrt{\dfrac{466675.1579}{18}} = 25926.3977$, and $\sigma^2 = 150^2 = 22500$.

Then $\chi^2 = \dfrac{(n-1)s^2}{\sigma^2} = \dfrac{18 \cdot 25926.3977}{22500} = 20.74$.

Step 7: Since χ^2 falls below χ^2_α we fail to reject H_0.

8. Step 1: H_0: $\pi = 50$ and H_1: $\pi > 50$.
Step 2: $\alpha = 0.01$.
Step 3: Since $np > 500$ and $n(100-p) > 500$, we use the z distribution.
Step 4: This is a one-tailed test. With $\alpha = 0.01$, the remaining area is $0.5000 - 0.0100 = 0.4900$ which corresponds to a z value of 2.33.
Step 5: Reject H_0 in favor of H_1 if $z > 2.33$. Otherwise, fail to reject H_0.
Step 6: Data: 48 out of 94. $p = \dfrac{48}{94} = 51.0638\%$ and $\sigma_p = \sqrt{\dfrac{\pi_0(100-\pi_0)}{n}} = \sqrt{\dfrac{50 \cdot 50}{94}} = 5.1571$.

$z = \dfrac{p-\pi_0}{\sigma_p} = \dfrac{51.0638 - 50}{5.1571} = 0.2063$.

Step 7: Since $z = 0.2063$ is below 2.33, we fail to reject H_0.

10. Step 1: H_0: $\mu = 2$, and H_1: $\mu \neq 2$.
Step 2: $\alpha = 0.01$.
Step 3: Since $n > 30$ we use the z distribution.
Step 4: This is a two-tailed test. With $\alpha = 0.01$, the remaining area is $0.5000 - 0.0050 = 0.4950$ which corresponds to a z value of 2.575.
Step 5: Reject H_0 in favor of H_1 if $z > 2.575$ or if $z < -2.575$. Otherwise, fail to reject H_0.
Step 6: Data: $\overline{x} = 1.79$; $s = 0.60$; $n = 216$. $\hat{\sigma}_{\overline{x}} = \dfrac{s}{\sqrt{n}} = \dfrac{0.60}{\sqrt{216}} = 0.0408$.

$z = \dfrac{\overline{x}-\mu_0}{\hat{\sigma}_{\overline{x}}} = \dfrac{1.79-2}{0.0408} = -5.147$.

Step 7: Since $z = -5.147 < -2.575$ we reject H_0.

12. Step 1: H_0: $\mu = \$2{,}150$ and H_1: $\mu > \$2{,}150$.
Step 2: $\alpha = 0.01$.
Step 3: Since $n < 30$ and we assume that population values are normally distributed, we use the t distribution.
Step 4: One-tailed test; with $\alpha = 0.01$ and $n = 7$ we have 6 df so we use $t_\alpha = t_{0.01} = 3.143$.
Step 5: Reject H_0 in favor of H_1 if $t > 3.143$. Otherwise, fail to reject H_0.
Step 6: Data: $n = 7$. $\overline{x} = \dfrac{16{,}493}{7} = \$2{,}356.14$. Calculate s.

x	$x-\overline{x}$	$(x-\overline{x})^2$
1,198	−1,158.14	1,341,288.2596
2,080	−276.14	76253.2996
1,130	−1,226.14	1,503,419.2996
1,510	−846.14	715952.8996
2,821	464.86	216094.8196
2,777	420.86	177123.1396
4,977	2,620.86	6,868,907.1396
16,493		10,899,038.8572

$s = \sqrt{\frac{\Sigma(x-\overline{x})^2}{n-1}} = \sqrt{\frac{10{,}899{,}038.8572}{6}} = 1{,}347.778.$ $\widehat{\sigma}_{\overline{x}} = \frac{s}{\sqrt{n}} = \frac{1{,}347.778}{\sqrt{7}} = 509.412$. Then
$t = \frac{\overline{x}-\mu_0}{\widehat{\sigma}_{\overline{x}}} = \frac{2{,}356.14 - 2{,}150}{509.412} = 0.4047.$
Step 7: Since $t < 3.143$ we fail to reject H_0.

14. Step 1: H_0: $\sigma = 2$ and H_1: $\sigma > 2$.
Step 2: $\alpha = 0.05$.
Step 3: Use χ^2 test on variances.
Step 4: One-tailed test, the acceptance region falls below χ^2_α. So with $\alpha = 0.05$ and 25 df we have $\chi^2_\alpha = \chi^2_{0.05} = 37.7$.
Step 5: Reject H_0 in favor of H_1 if $\chi^2 > 37.7$. Otherwise, fail to reject H_0.
Step 6: Since s = 3.26; $s^2 = 10.6276$ and $\sigma^2 = 4$. Then $\chi^2 = \frac{(n-1)s^2}{\sigma^2} = \frac{25 \cdot 10.6276}{4} = 66.4225$.
Step 7: Since $\chi^2 > 37.7$ we reject H_0.

16. Step 1: H_0: $\mu = \$1{,}000$ and H_1: $\mu \neq \$1{,}000$.
Step 2: $\alpha = 0.01$.
Step 3: Since $n > 30$ we use the z distribution.
Step 4: Two-tailed test with $\alpha = 0.01$ corresponds to a z value of 2.575.
Step 5: Reject H_0 in favor of H_1 if $|z| > \pm 2.575$.
Step 6: Data: $\overline{x} = \$981$; s = \$231; $n = 273$. $\widehat{\sigma}_{\overline{x}} = \frac{s}{\sqrt{n}} = \frac{231}{\sqrt{273}} = 13.98076$. Then
$z = \frac{981 - 1{,}000}{13.98076} = -1.359.$
Step 7: Since z is within ±2.575, we fail to reject H_0.

18. Step 1: H_0: $\mu = 27.37$ and H_1: $\mu \neq 27.37$.
Step 2: $\alpha = 0.01$.
Step 3: z distribution.
Step 4: Two-tailed test with $\alpha = 0.01$ corresponds to a z value of 2.575.
Step 5: Reject H_0 in favor of H_1 if $z < -2.575$ or if $z > 2.575$. Otherwise, fail to reject H_0.
Step 6: Data: $\overline{x} = 27.29$; $s = 3.75$; $n = 64$. $\widehat{\sigma}_{\overline{x}} = \frac{s}{\sqrt{n}} = \frac{3.75}{\sqrt{64}} = 0.46875$. Then
$z = \frac{27.29 - 27.37}{0.46875} = -0.1706.$
Step 7: Since z is within ±2.575, we fail to reject H_0.

20. Step 1: H_0: $\sigma^2 = 50{,}000$ and H_1: $\sigma^2 < 50{,}000$.
Step 2: $\alpha = 0.05$.
Step 3: Use χ^2 test on variances.

Step 4: One-tailed test, the rejection region falls below $\chi^2_{1-\alpha}$. So with $\alpha = 0.05$ and 10 df we have $\chi^2_{1-\alpha} = \chi^2_{0.95} = 3.94$.
Step 5: Reject H_0 in favor of H_1 if $\chi^2 < 3.94$. Otherwise, fail to reject H_0.
Step 6: Data: $n = 11$. Calculate $\overline{x}$ and s^2.

x	$x - \overline{x}$	$(x - \overline{x})^2$
1,200	−181.82	33058.5124
1,200	−181.82	33058.5124
1,000	−381.82	145786.5124
1,600	218.18	47602.5124
1,400	18.18	330.5124
1,400	18.18	330.5124
1,200	−181.82	33058.5124
1,700	318.18	101238.5124
1,600	218.18	47602.5124
1,300	−81.82	6694.5124
1,600	218.18	47602.5124
15,200		496363.6364

$\overline{x} = \dfrac{15{,}200}{11} = 1{,}381.82$ and $s^2 = \dfrac{\Sigma(x - \overline{x})^2}{n-1} = \dfrac{496,363.6364}{10} = 49,636.36364$. Then
$\chi^2 = \dfrac{(n-1)s^2}{\sigma^2} = \dfrac{10 \cdot 49,636.36364}{50,000} = 9.9273.$
Step 7: Since $\chi^2 > 3.94$ we fail to reject H_0.

22. Step 1: H_0: $\pi = 7$, and H_1: $\pi > 7$.
Step 2: $\alpha = 0.05$.
Step 3: z distribution.
Step 4: One-tailed test with $\alpha = 0.05$ corresponds to a z value of 1.645.
Step 5: Reject H_0 in favor of H_1 if $z > 1.645$. Otherwise, fail to reject H_0.
Step 6: Data: 139 out of 1,263, so $p = \dfrac{139}{1,263} \times (100 \text{ percent}) = 11.006\%$ and
$s_p = \dfrac{\pi_0(100 - \pi_0)}{n} = \sqrt{\dfrac{7 \cdot 93}{1,263}} = 0.7179\%$. Then $z = \dfrac{p - \pi_0}{s_p} = \dfrac{11.006 - 7}{0.7179} = 5.580.$
Step 7: Since $z > 1.645$ we reject H_0.

24. Step 1: H_0: $\mu = \$5{,}600$ and H_1: $\mu \neq \$5{,}600$.
Step 2: $\alpha = 0.05$.
Step 3: z distribution.
Step 4: Two-tailed test with $\alpha = 0.05$ corresponds to a z value of 1.96.
Step 5: Reject H_0 in favor of H_1 if $z < -1.96$ or if $z > 1.96$. Otherwise, fail to reject H_0.
Step 6: Data: $\overline{x} = \$5{,}750$; s = \$175; $n = 36$. $\widehat{\sigma}_{\overline{x}} = \dfrac{s}{\sqrt{n}} = \dfrac{175}{\sqrt{36}} = 29.1667$. Then
$z = \dfrac{5,750 - 5,600}{29.1667} = 5.1429.$
Step 7: Since $z > 1.96$, we reject H_0.

26. Step 1: H_0: $\mu = 50$ and H_1: $\mu \neq 50$.
Step 2: $\alpha = 0.01$.
Step 3: Although $n < 30$, we still use the z distribution since σ is known.
Step 4: This is a two-tailed test. With $\alpha = 0.01$, the remaining area is $0.5000 - 0.0050 = 0.4950$ which corresponds to a z value of -2.575.

Step 5: Reject H_0 in favor of H_1 if $z < -2.575$ or if $z > 2.575$. Otherwise, fail to reject H_0.

Step 6: Data: $\bar{x} = 49.25$; $s = 2$; $n = 24$. $\hat{\sigma}_{\bar{x}} = \frac{s}{\sqrt{n}} = \frac{2}{\sqrt{24}} = 0.4082$.

$$z = \frac{\bar{x} - \mu_0}{\hat{\sigma}_{\bar{x}}} = \frac{49.25 - 50}{0.4082} = -1.8371.$$

Step 7: Since z within ± 2.575, we fail to reject H_0.

28. Step 1: H_0: $\mu = 1{,}000$ and H_1: $\mu \neq 1{,}000$.

Step 2: $\alpha = 0.05$.

Step 3: t distribution.

Step 4: Two-tailed test with $\alpha = 0.05$ and 24 df we use $t_{\alpha/2} = t_{0.025} = 2.064$.

Step 5: Reject H_0 in favor of H_1 if $t < -2.064$ or if $t > 2.064$. Otherwise, fail to reject H_0.

Step 6: Data: $\bar{x} = 994$; s = 30; $n = 25$. $\hat{\sigma}_{\bar{x}} = \frac{s}{\sqrt{n}} = \frac{30}{\sqrt{25}} = 6$. Then $t = \frac{994 - 1{,}000}{6} = -1$.

Step 7: Since t is within ± 2.064 we fail to reject H_0.

30. Step 1: H_0: $\mu = 50$ and H_1: $\mu \neq 50$.

Step 2: $\alpha = 0.05$.

Step 3: t distribution.

Step 4: Two-tailed test with $\alpha = 0.05$ and 9 df we use $t_{\alpha/2} = t_{0.025} = 2.262$.

Step 5: Reject H_0 in favor of H_1 if $t < -2.262$ or if $t > 2.262$. Otherwise, fail to reject H_0.

Step 6: Data: $\bar{x} = 50.02$; s = 0.024; $n = 10$. $\hat{\sigma}_{\bar{x}} = \frac{s}{\sqrt{n}} = \frac{0.024}{\sqrt{10}} = 0.007589$. Then

$$t = \frac{50.02 - 50}{0.007589} = 2.635.$$

Step 7: Since $t > 2.262$ we reject H_0.

32. Step 1: H_0: $\mu = 75$ and H_1: $\mu > 75$.

Step 2: $\alpha = 0.01$.

Step 3: t distribution.

Step 4: One-tailed test with $\alpha = 0.01$ and 8 df we use $t_\alpha = t_{0.01} = 2.896$.

Step 5: Reject H_0 in favor of H_1 if $t > 2.896$. Otherwise, fail to reject H_0.

Step 6: Data: $\bar{x} = 80$; s = 4; $n = 9$. $\hat{\sigma}_{\bar{x}} = \frac{s}{\sqrt{n}} = \frac{4}{\sqrt{9}} = 1.3333$. Then $t = \frac{80 - 75}{1.3333} = 3.75$.

Step 7: Since $t > 2.896$ we reject H_0.

34. Step 1: H_0: $\mu = 1.5$ and H_1: $\mu \neq 1.5$.

Step 2: $\alpha = 0.05$.

Step 3: Since we know the population standard deviation we use the z distribution.

Step 4: Two-tailed test with $\alpha = 0.05$ corresponds to a z value of 1.96.

Step 5: Reject H_0 in favor of H_1 if $z > 1.96$ or if $z < -1.96$. Otherwise, fail to reject H_0.

Step 6: Data: $\bar{x} = 1.5005$; s = 0.01; $n = 20$. $\hat{\sigma}_{\bar{x}} = \frac{s}{\sqrt{n}} = \frac{0.01}{\sqrt{20}} = 0.002236$. Then

$$z = \frac{1.5005 - 1.5}{0.002236} = 0.2236.$$

Step 7: Since $z > -1.96$ we fail to reject H_0.

36. Step 1: H_0: $\mu = 4{,}000$ and H_1: $\mu < 4{,}000$.

Step 2: $\alpha = 0.05$.

Step 3: t distribution.

Step 4: One-tailed test with $\alpha = 0.05$ and 12 df we use $t_\alpha = t_{0.05} = -1.782$.
Step 5: Reject H_0 in favor of H_1 if $t < -1.782$. Otherwise, fail to reject H_0.
Step 6: Data: $\overline{x} = 3,950$; s = 100; $n = 13$. $\widehat{\sigma}_{\overline{x}} = \frac{s}{\sqrt{n}} = \frac{100}{\sqrt{13}} = 27.7350$. Then

$$t = \frac{3,950 - 4,000}{27.7350} = -1.8027.$$

Step 7: Since $t < -1.782$, we reject H_0. The shipment should be rejected.

38. Step 1: H_0: $\mu = \$10{,}000$ and H_1: $\mu > \$10{,}000$.
Step 2: $\alpha = 0.05$.
Step 3: t distribution.
Step 4: One-tailed test with $\alpha = 0.05$ and 14 df we use $t_\alpha = t_{0.05} = 1.761$.
Step 5: Reject H_0 in favor of H_1 if $t > 1.761$. Otherwise, fail to reject H_0.
Step 6: Data: $\overline{x} = \$10{,}575$; s = \$600; $n = 15$. $\widehat{\sigma}_{\overline{x}} = \frac{s}{\sqrt{n}} = \frac{600}{\sqrt{15}} = 154.9193$. Then

$$t = \frac{10,575 - 10,000}{154.9193} = 3.7116.$$

Step 7: Since $t > 1.761$, we reject H_0, there is sufficient evidence that the contractor averages more than \$10,000 per consultation.

40. Step 1: H_0: $\mu = 7000$ and H_1: $\mu < 7000$.
Step 2: $\alpha = 0.05$.
Step 3: t distribution.
Step 4: One-tailed test with $\alpha = 0.05$ and 5 df we use $t_\alpha = t_{0.05} = 2.015$.
Step 5: Reject H_0 in favor of H_1 if $t < -2.015$. Otherwise, fail to reject H_0.
Step 6: Data: $\overline{x} = \frac{43281}{6} = 7213.5$; $\sigma = 875$; $n = 6$. $\sigma_{\overline{x}} = \frac{\sigma}{\sqrt{n}} = \frac{875}{\sqrt{6}} = 357.217$. Then

$$t = \frac{7213.5 - 7000}{357.217} = 0.598.$$

Step 7: Since $t > -2.015$, we fail to reject H_0.

42. Step 1: H_0: $\pi = 83.1$, and H_1: $\pi \neq 83.1$.
Step 2: $\alpha = 0.05$.
Step 3: z distribution.
Step 4: Two-tailed test with $\alpha = 0.05$ corresponds to a z value of 1.96.
Step 5: Reject H_0 in favor of H_1 if $z > 1.96$ or if $z < -1.96$. Otherwise, fail to reject H_0.
Step 6: Data: 113 out of 157, so $p = \frac{113}{157} = 71.97\%$ and $s_p = \frac{\pi_0(100 - \pi_0)}{n} = \sqrt{\frac{83.1 \cdot 16.9}{157}} = 2.9908\%$.

Then $z = \frac{p - \pi_0}{s_p} = \frac{71.97 - 83.1}{2.9908} = -3.721$.

Step 7: Since $z < -1.96$ we reject H_0.

44. Step 1: H_0: $\sigma = 5000$ and H_1: $\sigma > 5000$.
Step 2: $\alpha = 0.05$.
Step 3: Use χ^2 test on variances.
Step 4: One-tailed test, the rejection region falls above χ^2_α. So with $\alpha = 0.05$ and 21 df we have $\chi^2_\alpha = \chi^2_{0.05} = 32.7$.
Step 5: Reject H_0 in favor of H_1 if $\chi^2 > 32.7$. Otherwise, fail to reject H_0.
Step 6: Data: $n = 22$ and from MINITAB, $s = 7912$. So $\chi^2 = \frac{(n-1)s^2}{\sigma^2} = \frac{21 \cdot (7912)^2}{(5000)^2} = 52.583$.
Step 7: Since $\chi^2 > 32.7$, we reject H_0, and conclude the standard deviation is greater than 5000.

46. Step 1: H_0: $\sigma = 30$ mg/dl and H_1: $\sigma \neq 30$ mg/dl.
Step 2: $\alpha = 0.05$.
Step 3: Use χ^2 test on variances.
Step 4: This is a two-tailed test, the rejection region falls above $\chi^2_{\alpha/2}$ and below $\chi^2_{1-\alpha/2}$. With $\alpha = 0.05$ and $n = 19$ we have 18 df so $\chi^2_{\alpha/2} = \chi^2_{0.025} = 31.5$, while $\chi^2_{1-\alpha/2} = \chi^2_{0.975} = 8.23$.
Step 5: Reject H_0 in favor of H_1 if $\chi^2 > 31.5$ or if $\chi^2 < 8.23$. Otherwise, fail to reject H_0.
Step 6: Data: $n = 19$ and from MINITAB, $s = 51.7$. So $\chi^2 = \frac{(n-1)s^2}{\sigma^2} = \frac{18 \cdot (51.7)^2}{(30)^2} = 53.458$.
Step 7: Since $\chi^2 > 31.5$, we reject H_0, the standard deviation is not 30 mg/dl.

48. Step 1: H_0: $\pi = 4.1$ and H_1: $\pi \neq 4.1$.
Step 2: $\alpha = 0.10$.
Step 3: Since $np > 500$ and $n(100 - p) > 500$, we use the z distribution.
Step 4: This is a two-tailed test. With $\alpha = 0.10$, the remaining area in each tail is $0.5000 - 0.0500 = 0.4500$ which corresponds to a z value of 1.645.
Step 5: Reject H_0 in favor of H_1 if $z < -1.645$ or if $z > 1.645$. Otherwise, fail to reject H_0.
Step 6: Data: 19 out of 150. $p = \frac{19}{150} \times (100 \text{ percent}) = 12.67\%$ and

$$\sigma_p = \sqrt{\frac{\pi_0(100 - \pi_0)}{n}} = \sqrt{\frac{4.1 \cdot 95.9}{150}} = 1.6190. \quad z = \frac{p - \pi_0}{\sigma_p} = \frac{12.67 - 4.1}{1.6190} = 5.29.$$

Step 7: Since $z > 1.645$, we reject H_0, the percent of African Americans at San Marcos has changed.

50. Step 1: H_0: $\mu = 3500$ psi and H_1: $\mu \neq 3500$ psi.
Step 2: $\alpha = 0.05$.
Step 3: t distribution.
Step 4: Two-tailed test with $\alpha = 0.05$ and 13 df we use $t_{\alpha/2} = t_{0.025} = 2.160$.
Step 5: Reject H_0 in favor of H_1 if $t > 2.160$ or if $t < -2.160$. Otherwise, fail to reject H_0.
Step 6: Data: from MINITAB: $\overline{x} = 3496.9$, $s = 202.6$; $n = 14$. So $\hat{\sigma}_{\overline{x}} = \frac{s}{\sqrt{n}} = \frac{202.6}{\sqrt{14}} = 54.147$. Then

$$t = \frac{3496.9 - 3500}{54.147} = -0.057.$$

Step 7: Since $-2.160 < t < 2.160$, we fail to reject H_0.

Chapter 9 Inference: Two-Sample Procedures

Answers to Even-Numbered Self-Testing Review

9-1 Hypothesis Tests of Two Variances

2. Step 1: State the null and alternative hypotheses. H_0: $\sigma_1^2 = \sigma_2^2$ and H_1: $\sigma_1^2 > \sigma_2^2$.
Step 2: Select the level of significance. Use $\alpha = 0.01$.
Step 3: Determine the test distribution to use. We use an F distribution. Since this is a right-tailed test, we use the table with 0.01 in the right tail. The larger variance has a sample size of 13, so the df for the numerator is $13 - 1$ or 12. The sample size for the smaller variance is 21, so $21 - 1$ or 20 is the df for the denominator.
Step 4: Define the critical or rejection region(s). You look across the top line of the F table for a df value of 12 and down the left column until you get to the line with df $= 20$. The critical F value is 3.23.
Step 5: State the decision rule. Reject H_0 in favor of H_1 if $\boldsymbol{F} > 3.23$. Otherwise, fail to reject H_0.
Step 6: Compute the Test Statistic. The $\boldsymbol{F} = \dfrac{48.91}{35.43} = 1.380$.
Step 7: Make the statistical decision. Since a $\boldsymbol{F} = 1.380$ does not fall in the rejection region, we fail to reject H_0 that the variances of both populations are equal.

4. Step 1: Although this problems reads as though it might be a left-tailed test, remember that the sample with the greater variance is always designated as sample 1. So we designate the Shop N Pay sample as sample 1 and claim its variance is greater than the variance of Buy Fair. Thus, H_0: $\sigma_1^2 = \sigma_2^2$ and H_1: $\sigma_1^2 > \sigma_2^2$.
Step 2: Use $\alpha = 0.05$.
Step 3: We use an F distribution. Since this is a right-tailed test, we use the table with 0.05 in the right tail. The df for the numerator is $7 - 1$ or 6. And the df for the denominator is $25 - 1$ or 24.
Step 4: The critical F value is 2.51.
Step 5: Reject H_0 in favor of H_1 if $\boldsymbol{F} > \mathbf{2.51}$. Otherwise, fail to reject H_0.
Step 6: The $\boldsymbol{F} = \dfrac{17.84}{15.93} = 1.120$
Step 7: Since a $\boldsymbol{F}$ of 1.120 does not fall in the rejection region, we fail to reject H_0 that the variances are equal for both stores.

6. Step 1: H_0: $\sigma_1^2 = \sigma_2^2$ and H_1: $\sigma_1^2 \neq \sigma_2^2$.
Step 2: Use $\alpha = 0.05$.
Step 3: We use an F distribution. Since this is a two-tailed test, we use the table with 0.025 in the right tail. The df for the numerator is $6 - 1 = 5$. And the df for the denominator is $7 - 1 = 6$.
Step 4: The critical F value is 5.99.
Step 5: Reject H_0 in favor of H_1 if $\boldsymbol{F} > 5.99$. Otherwise, fail to reject H_0.
Step 6: Then $\boldsymbol{F} = \dfrac{0.1024}{0.0064} = 16$.

Step 7: Since $F = 16$, we reject H_0 in favor of H_1, the variances are not equal.

8. Step 1: H_0: $\sigma_1^2 = \sigma_2^2$ and H_1: $\sigma_1^2 > \sigma_2^2$.
Step 2: Use $\alpha = 0.05$.
Step 3: We use an F distribution. Since this is a right-tailed test, we use the table with 0.05 in the right tail. The df for the numerator is $16 - 1$ or 15. And the df for the denominator is $28 - 1$ or 27.
Step 4: The critical F value is 2.06.
Step 5: Reject H_0 in favor of H_1 if $F > 2.06$. Otherwise, fail to reject H_0.
Step 6: The $F = \dfrac{4.9}{1.3} = 3.769$.
Step 7: Since $F = 3.769$ falls in the rejection region, we reject H_0 in favor of the H_1 that the machine set at a faster speed had a greater variance in output.

10. Step 1: H_0: $\sigma_1^2 = \sigma_2^2$ and H_1: $\sigma_1^2 \neq \sigma_2^2$.
Step 2: Use $\alpha = 0.10$.
Step 3: We use an F distribution. Since this is a two-tailed test, we use the table with 0.05 in the right tail. The df for both the numerator and the denominator is $20 - 1$ or 19.
Step 4: The critical F value is 2.16.
Step 5: Reject H_0 in favor of H_1 if $F > 2.16$. Otherwise, fail to reject H_0.
Step 6: Using MINITAB, we find for Polonio Pass, $s = 527$; and for Devil's Den, $s = 578$. Then
$F = \dfrac{578^2}{527^2} = \dfrac{334084}{277729} = 1.203$.
Step 7: Since $F = 1.203 < 2.16$, we fail to reject H_0 that the variances are equal for both pumping plants.

12. Step 1: H_0: $\sigma_1^2 = \sigma_2^2$ and H_1: $\sigma_1^2 > \sigma_2^2$.
Step 2: Use $\alpha = 0.05$.
Step 3: We use an F distribution. Since this is a right-tailed test, we use the table with 0.05 in the right tail. The df for both the numerator and the denominator is $3 - 1$ or 2.
Step 4: The critical F value is 19.00.
Step 5: Reject H_0 in favor of H_1 if $F > 19.00$. Otherwise, fail to reject H_0.
Step 6: Using MINITAB, we find for Sugarcane field $s = 40953$ while for Pasture $s = 7569$. Thus
$F = \dfrac{40953^2}{7569^2} = \dfrac{1677148209}{57289761} = 29.275$.
Step 7: Since $F = 29.275$ falls in the rejection region, we reject H_0 in favor of the H_1 that the sugarcane fields had a greater variance in biomass.

14. Step 1: H_0: $\sigma_1^2 = \sigma_2^2$ and H_1: $\sigma_1^2 \neq \sigma_2^2$.
Step 2: Use $\alpha = 0.10$.
Step 3: We use an F distribution. Since this is a two-tailed test, we use the table with 0.05 in the right tail. Using MINITAB, we find for the standard method we have $n_1 = 10$ and $s_1 = 283.6$; for the megasonic we have $n_2 = 5$ and $s_2 = 210.5$ The df for the numerator is $10 - 1$ or 9. And the df for the denominator is $5 - 1$ or 4
Step 4: The critical F value is 6.00.
Step 5: Reject H_0 in favor of H_1 if $F > 6.00$. Otherwise, fail to reject H_0.
Step 6: Then $F = \dfrac{283.6^2}{210.5^2} = \dfrac{80428.96}{44310.25} = 1.815$.
Step 7: Since $F < 6.00$, we fail to reject H_0 that the variances are equal for both methods.
We are assuming that the population of the number of defects is normally distributed.

9-2 Inference About Two Means

2. Step 1: H_0: $\mu_1 = \mu_2$ and H_1: $\mu_1 > \mu_2$.
Step 2: $\alpha = 0.05$.
Step 3: Since the samples are independent and both samples are large, we use procedure 2 with the z distribution.
Step 4: The critical z value is 1.645.
Step 5: Reject H_0 in favor of H_1 if $z > 1.645$. Otherwise, fail to reject H_0.
Step 6: Then $z = \dfrac{\overline{x}_1 - \overline{x}_2}{\sqrt{\dfrac{s_1^2}{n_1} + \dfrac{s_2^2}{n_2}}} = \dfrac{24.1 - 23.9}{\sqrt{\dfrac{5.24^2}{100} + \dfrac{6.29^2}{73}}} = 0.2248.$
Step 7: Since $z = 0.2248$ does not fall in the rejection region, we fail to reject the null hypothesis. The physician-led group did not do significantly better than the control group.

4. Step 1: H_0: $\mu_1 = \mu_2$ and H_1: $\mu_1 > \mu_2$.
Step 2: $\alpha = 0.01$.
Step 3: Since the samples are independent and both samples are large, we use procedure 2 with the z distribution.
Step 4: The critical z value is 2.33.
Step 5: Reject H_0 in favor of H_1 if $z > 2.33$. Otherwise, fail to reject H_0.
Step 6: Then $z = \dfrac{\overline{x}_1 - \overline{x}_2}{\sqrt{\dfrac{s_1^2}{n_1} + \dfrac{s_2^2}{n_2}}} = \dfrac{38.08 - 35.60}{\sqrt{\dfrac{0.84^2}{44} + \dfrac{0.79^2}{47}}} = \dfrac{2.48}{0.1712} = 14.49.$
Step 7: Since $z = 14.49$ falls in the rejection region, we reject H_0. The PD at Nordstrom's is greater than the PD at the May Company.

6. Step 1: H_0: $\mu_1 = \mu_2$ and H_1: $\mu_1 > \mu_2$.
Step 2: $\alpha = 0.05$.
Step 3: Since the samples are independent and both samples are small, we first check to see if the populations have equal variances. Since $F = \dfrac{0.8050}{0.4377} = 1.839$, there is insufficient evidence to show that the variances differ, so we use a t distribution and procedure 4.
Step 4: The critical t value with 18 df is 1.734.
Step 5: Reject H_0 in favor of H_1 if $t > 1.734$. Otherwise, fail to reject H_0.
Step 6: Then $t = \dfrac{\overline{x}_1 - \overline{x}_2}{s_p\sqrt{\dfrac{1}{n_1} + \dfrac{1}{n_2}}} = \dfrac{5.853 - 6.509}{(0.768)\sqrt{\dfrac{1}{10} + \dfrac{1}{10}}} = -1.67.$
Step 7: Since $t = -1.67$ does not fall in the rejection region, we fail to reject H_0. The ratings appear equal.

8. To obtain a 95% confidence interval for $\mu_1 - \mu_2$, we need $z_{\alpha/2} = z_{0.025} = 1.96$. Then we compute:

$$\overline{x}_1 - \overline{x}_2 \pm z_{\alpha/2}\sqrt{\frac{\sigma_1^2}{n_1} + \frac{\sigma_2^2}{n_2}} = (2.84 - 2.21) \pm (1.96)\sqrt{\frac{1.10^2}{32} + \frac{0.82^2}{35}} = 0.63 \pm (1.96)(0.239)$$
$$= 0.63 \pm 0.468 \text{ or } 0.162 \text{ to } 1.098.$$

10. To obtain a 95% confidence interval for $\mu_1 - \mu_2$, we need $z_{\alpha/2} = z_{0.025} = 1.96$. Then we compute:

$$\overline{x}_1 - \overline{x}_2 \pm z_{\alpha/2}\sqrt{\frac{\sigma_1^2}{n_1} + \frac{\sigma_2^2}{n_2}} = (42.0 - 37.2) \pm (1.96)\sqrt{\frac{34.7^2}{141} + \frac{33.4^2}{123}} = 4.8 \pm (1.96)(4.196)$$
$$= 4.8 \pm 8.22 \text{ or } -3.42 \text{ to } 13.02.$$

12. Step 1: $\sigma_1^2 = \sigma_2^2$ and H_1: $\sigma_1^2 \neq \sigma_2^2$.

Step 2: Use $\alpha = 0.02$.

Step 3: We use an F distribution. Since this is a two-tailed test, we use the table with 0.01 in the right tail. The df for the numerator is $5-1$ or 4. And the df for the denominator is $5-1$ or 4.

Step 4: The critical F value is 15.98.

Step 5: Reject H_0 in favor of H_1 if $\boldsymbol{F} > 15.98$. Otherwise, fail to reject H_0.

Step 6: The $\boldsymbol{F} = \dfrac{160^2}{85^2} = 3.543$.

Step 7: Since $\boldsymbol{F} = 3.543$, we fail to reject H_0.

We have independent populations with $\sigma_1^2 = \sigma_2^2$ (or at least not $\sigma_1^2 \neq \sigma_2^2$) and sample size of both samples is less than 30. The only assumption we must make is that these samples are taken from normal populations with unknown but equal variances. To test H_0: $\mu_1 = \mu_2$, the test statistic is:

$$t = \frac{\overline{x}_1 - \overline{x}_2}{s_p\sqrt{\frac{1}{n_1}+\frac{1}{n_2}}} = \frac{\overline{x}_1 - \overline{x}_2}{\sqrt{\frac{s_1^2(n_1-1)+s_2^2(n_2-1)}{n_1+n_2-2}}\sqrt{\frac{1}{n_1}+\frac{1}{n_2}}}.$$

This has a t distribution with n_1+n_2-2 degrees of freedom.

Step 1: H_0: $\mu_1 = \mu_2$ and H_1: $\mu_1 \neq \mu_2$.

Step 2: $\alpha = 0.01$.

Step 3: Since the samples are independent and both samples are small, we already tested to see if the variances are equal and we have reason to believe that. This means procedure 4 and a t test is in order. The df $= 5+5-2 = 8$.

Step 4: With df $= 8$, $t_{0.005} = 3.355$.

Step 5: Reject H_0 in favor of H_1 if $\boldsymbol{t} > 3.355$ or if $\boldsymbol{t} < -3.355$. Otherwise, fail to reject H_0.

Step 6: Then

$$\boldsymbol{t} = \frac{\overline{x}_1 - \overline{x}_2}{\sqrt{\frac{s_1^2(n_1-1)+s_2^2(n_2-1)}{n_1+n_2-2}}\sqrt{\frac{1}{n_1}+\frac{1}{n_2}}} = \frac{440-340}{\sqrt{\frac{(160)^2(5-1)+(85)^2(5-1)}{5+5-2}}\sqrt{\frac{1}{5}+\frac{1}{5}}}$$

$$= \frac{100}{\sqrt{164125}\sqrt{0.4}} = 1.234.$$

Step 7: Since $\boldsymbol{t}$ does not fall in the rejection region, we fail to reject H_0.

14. Since the samples are independent and both samples are small, we first test to see if the variances are equal.

Step 1: H_0: $\sigma_1^2 = \sigma_2^2$ and H_1: $\sigma_1^2 \neq \sigma_2^2$.

Step 2: Use $\alpha = 0.05$.

Step 3: We use an F distribution. Since this is a two-tailed test, we use the table with 0.025 in the right tail. The df for the numerator is $25-1 = 24$. And the df for the denominator is $25-1 = 24$.

Step 4: The critical F value is 2.27.

Step 5: Reject H_0 in favor of H_1 if $\boldsymbol{F} > 2.27$. Otherwise, fail to reject H_0.

Step 6: Then $\boldsymbol{F} = \dfrac{16.8^2}{11.6^2} = \dfrac{282.24}{134.56} = 2.098$.

Step 7: Since $\boldsymbol{F} < 2.27$, we fail to reject H_0 that the variances are equal for both methods.

We now test whether the means are equal.

Step 1: H_0: $\mu_1 = \mu_2$ and H_1: $\mu_1 \neq \mu_2$.

Step 2: $\alpha = 0.01$.

Step 3: The samples are independent and both samples are small, and from the hypothesis test of equal variance we have reason to believe that the variances are equal. This means procedure 4 and a t test is in order. The df $= 25+25-2 = 48$.

Step 4: There is no t value for 48 df in Appendix 4, but we can use the closest value (on the 40 df line) to get $t = 2.704$.
Step 5: Reject H_0 in favor of H_1 if $t < -2.704$ or if $t > 2.704$. Otherwise, fail to reject H_0.
Step 6: Then

$$t = \frac{\overline{x}_1 - \overline{x}_2}{\sqrt{\frac{s_1^2(n_1-1)+s_2^2(n_2-1)}{n_1+n_2-2}}\sqrt{\frac{1}{n_1}+\frac{1}{n_2}}} = \frac{93.2 - 107.8}{\sqrt{\frac{16.8^2(24)+11.6^2(24)}{25+25-2}}\sqrt{\frac{1}{25}+\frac{1}{25}}} = \frac{-6.20}{3.6056}$$

$= -3.576$.

Step 7: Since a $t = -3.576$ falls in the rejection region, we reject H_0. There appears to be a significant difference between the IQ's of schizophrenic patients and those of normal subjects.

16. To obtain a 95% confidence interval for μ_d, with df = 9, we need $t_{\alpha/2} = t_{0.025} = 2.262$. From Exercise 15 $\overline{d} = 8.2$ and $\frac{s_d}{\sqrt{n}} = 5.731$. We then compute: $\overline{d} \pm t_{\alpha/2}\frac{s_d}{\sqrt{n}} = 8.2 \pm (2.262)(5.731) = 8.2 \pm 12.96$ or -4.76 to 21.16.

18. Step 1: H_0: $\mu_1 = \mu_2$ and H_1: $\mu_1 > \mu_2$.
Step 2: $\alpha = 0.05$.
Step 3: Since the samples are independent and both samples are small, we should first test to see if the variances are equal. Based on this test, we have reason to believe that they are. This means Procedure 4 and a t test is in order. The df $= 13 + 10 - 2 = 21$.
Step 4: The critical value is $t = 1.721$.
Step 5: Reject H_0 in favor of H_1 if $t > 1.721$. Otherwise, fail to reject H_0.
Step 6: Then

$$t = \frac{\overline{x}_1 - \overline{x}_2}{\sqrt{\frac{s_1^2(n_1-1)+s_2^2(n_2-1)}{n_1+n_2-2}}\sqrt{\frac{1}{n_1}+\frac{1}{n_2}}} = \frac{116 - 109}{\sqrt{\frac{23^2(12)+13^2(9)}{13+10-2}}\sqrt{\frac{1}{13}+\frac{1}{10}}} = \frac{7}{8.1422} = 0.860.$$

Step 7: Since $t = 0.860$ does not fall in the rejection region, we fail to reject H_0. There appears to be no significant difference in the mean IQ scores.

20. Step 1: H_0: $\mu_1 = \mu_2$ and H_1: $\mu_1 \neq \mu_2$.
Step 2: $\alpha = 0.01$.
Step 3: Since the samples are independent and both samples are small, we should first test to see if the variances are equal. Based on this test, we have reason to believe that they are. This means procedure 4 and a t test is in order. The df $= 5 + 5 - 2 = 8$.
Step 4: The critical values of $t = \pm 3.355$.
Step 5: Reject H_0 in favor of H_1 if $t < -3.355$ or $t > +3.355$. Otherwise, fail to reject H_0.
Step 6: Then

$$t = \frac{\overline{x}_1 - \overline{x}_2}{\sqrt{\frac{s_1^2(n_1-1)+s_2^2(n_2-1)}{n_1+n_2-2}}\sqrt{\frac{1}{n_1}+\frac{1}{n_2}}} = \frac{40.7 - 34.5}{\sqrt{\frac{7^2(4)+4^2(4)}{5+5-2}}\sqrt{\frac{1}{5}+\frac{1}{5}}} = \frac{6.20}{3.6056} = 1.720.$$

Step 7: Since $t = 1.720$ does not fall in the rejection region, we fail to reject H_0. We cannot conclude that there's a significant difference in metabolic rates measured at the thalamus.

22. Step 1: H_0: $\mu_1 = \mu_2$ and H_1: $\mu_1 \neq \mu_2$.
Step 2: $\alpha = 0.05$.
Step 3: Since the samples are independent and both samples are small, we should first test to see if the variances are equal. However, we have reason to believe that they are. This means procedure 4 and a t test is in order. The df $= 10 + 10 - 2 = 18$.

Step 4: The critical values of $t = \pm 2.101$.
Step 5: Reject H_0 in favor of H_1 if $t < -2.101$ or $t > +2.101$. Otherwise, fail to reject H_0.
Step 6: Then

$$t = \frac{\overline{x}_1 - \overline{x}_2}{\sqrt{\frac{s_1^2(n_1-1)+s_2^2(n_2-1)}{n_1+n_2-2}}\sqrt{\frac{1}{n_1}+\frac{1}{n_2}}} = \frac{0.32-0.41}{\sqrt{\frac{(0.00027)^2(9)+(0.00018)^2(9)}{10+10-2}}\sqrt{\frac{1}{10}+\frac{1}{10}}}$$

$$= \frac{-0.09}{0.0001026} = -877.2.$$

Step 7: Since $t = -877.2$ falls in the rejection region, we reject H_0. There is a highly significant difference between the two sides in the duration of single-limb support for amputees.

24. Step 1: H_0: $\mu_d = 0$ and H_1: $\mu_d \neq 0$.
Step 2: $\alpha = 0.01$.
Step 3: We use a paired t test and a t distribution with 19 df
Step 4: The critical values of $t = \pm 2.861$.
Step 5: Reject H_0 in favor of H_1 if $t < -2.861$ or $t > +2.861$. Otherwise, fail to reject H_0.
Step 6: Then $t = \dfrac{\overline{d} - \mu_d}{\frac{s_d}{\sqrt{n}}} = \dfrac{-0.374}{0.161} = -2.323$.
Step 7: Since $t = -2.323$ does not fall in the rejection region, we fail to reject H_0. The rating aren't significantly different.

26. Step 1: H_0: $\mu_1 = \mu_2$ and H_1: $\mu_1 \neq \mu_2$.
Step 2: $\alpha = 0.01$.
Step 3: Since the samples are independent and both samples are small, we first check to see if the populations have equal variances. We use a t distribution and procedure 3.
Step 4: The critical t values with $6 - 1$ or 5 df is ± 4.032.
Step 5: Reject H_0 in favor of H_1 if $t < -4.032$ or $t > +4.032$. Otherwise, fail to reject H_0.
Step 6: Then $t = \dfrac{\overline{x}_1 - \overline{x}_2}{\sqrt{\frac{s_1^2}{n_1}+\frac{s_2^2}{n_2}}} = \dfrac{4.17-4.00}{\sqrt{\frac{0.32^2}{6}+\frac{0.08^2}{7}}} = 1.268$.
Step 7: Since $t = 1.268$, we fail to reject H_0 that there is no significant difference on the intention to buy scale.

28. To obtain a 95% confidence interval for $\mu_1 - \mu_2$, we need $z_{\alpha/2} = z_{0.025} = 1.96$. Then we compute:

$$\overline{x}_1 - \overline{x}_2 \pm z_{\alpha/2}\sqrt{\frac{\sigma_1^2}{n_1}+\frac{\sigma_2^2}{n_2}} = (73,000 - 89,000) \pm (1.96)\sqrt{\frac{11,000^2}{37}+\frac{21,000^2}{52}}$$
$$= -16,000 \pm (1.96)(3,428) = -16,000 \pm 6,718.88$$

or $-\$22,718.88$ to $-\$9,281.12$.

30. For all seven tests:
Step 1: H_0: $\mu_1 = \mu_2$ and H_1: $\mu_1 \neq \mu_2$.
Step 2: $\alpha = 0.05$.
Step 3: Based on the results of Exercise 29, we use procedure 4 with the t distribution with 58 df.
Step 4: The critical t value is the conservative 2.021 (40 df).
Step 5: Reject H_0 in favor of H_1 if $t < -2.021$ or $t > 2.021$. Otherwise, fail to reject H_0.

Step 6: We calculate the test statistic and make the decision in for variable in the following table. Here the test statistic is $t = \dfrac{\overline{x}_1 - \overline{x}_2}{s_p\sqrt{\dfrac{1}{n_1} + \dfrac{1}{n_2}}}$.

Variable	Calculate z	Step 7: Make the decision.
Quality	$t = \dfrac{4.10 - 3.73}{(0.869)\sqrt{\dfrac{1}{30} + \dfrac{1}{30}}} = 1.649$	Fail to reject H_0.
Reliability	$t = \dfrac{4.27 - 3.73}{(1.005)\sqrt{\dfrac{1}{30} + \dfrac{1}{30}}} = 2.081$	Reject H_0.
Reputation	$t = \dfrac{4.67 - 3.67}{(0.794)\sqrt{\dfrac{1}{30} + \dfrac{1}{30}}} = 4.878$	Reject H_0.
Value	$t = \dfrac{3.80 - 3.87}{(0.979)\sqrt{\dfrac{1}{30} + \dfrac{1}{30}}} = -0.277$	Fail to reject H_0.
Worthwhile Buy	$t = \dfrac{4.03 - 3.73}{(1.114)\sqrt{\dfrac{1}{30} + \dfrac{1}{30}}} = 1.043$	Fail to reject H_0.
Reasonable Buy	$t = \dfrac{3.33 - 3.90}{(1.074)\sqrt{\dfrac{1}{30} + \dfrac{1}{30}}} = -2.056$	Reject H_0.
Intend to Buy	$t = \dfrac{3.97 - 3.53}{(1.125)\sqrt{\dfrac{1}{30} + \dfrac{1}{30}}} = 1.514$	Fail to reject H_0.

32. For all seven tests:
Step 1: H_0: $\sigma_1^2 = \sigma_2^2$ and H_1: $\sigma_1^2 \neq \sigma_2^2$.
Step 2: Use $\alpha = 0.05$.
Step 3: We use an F distribution. Since this is a two-tailed test, we use the table with 0.025 in the right tail. The df for the numerator and the denominator is $27 - 1 = 26$. We will use the value from the table with df of the denominator is 24.
Step 4: The critical F value is 2.22.
Step 5: Reject H_0 in favor of H_1 if $F > 2.22$. Otherwise, fail to reject H_0.
Step 6: Then $F = \dfrac{s_1^2}{s_2^2}$, with $s_1 \geq s_2$. We calculate the test statistic and make the decision in for variable in the following table.

Variable	Calculate F	Step 7: Make the decision.
Quality	$F = \dfrac{0.97^2}{0.62^2} = 2.297$	Reject H_0, $\sigma_1^2 \neq \sigma_2^2$.
Reliability	$F = \dfrac{0.74^2}{0.58^2} = 1.628$	Fail to reject H_0, $\sigma_1^2 = \sigma_2^2$.
Reputation	$F = \dfrac{0.80^2}{0.22^2} = 3.628$	Reject H_0, $\sigma_1^2 \neq \sigma_2^2$.

Variable	Calculate F	Step 7: Make the decision.
Value	$F = \dfrac{1.02^2}{0.64^2} = 2.540$	Reject H_0, $\sigma_1^2 \neq \sigma_2^2$.
Worthwhile Buy	$F = \dfrac{0.78^2}{0.72^2} = 1.174$	Fail to reject H_0, $\sigma_1^2 = \sigma_2^2$.
Reasonable Buy	$F = \dfrac{0.94^2}{0.75^2} = 1.571$	Fail to reject H_0, $\sigma_1^2 = \sigma_2^2$.
Intend to Buy	$F = \dfrac{1.03^2}{0.83^2} = 1.540$	Fail to reject H_0, $\sigma_1^2 = \sigma_2^2$.

We now test the hypothesis that the population means are equal. For all seven tests:
Step 1: H_0: $\mu_1 = \mu_2$ and H_1: $\mu_1 \neq \mu_2$.
Step 2: $\alpha = 0.05$.
Step 3: If we rejected H_0 in the previous test of variance, then $\sigma_1^2 \neq \sigma_2^2$, so we use procedure 3 and a t distribution with $27 - 1 = 26$ df. If we failed to reject H_0 in the previous test of variance, then $\sigma_1^2 = \sigma_2^2$, so we use procedure 4 and a t distribution with $27 + 27 - 2 = 52$ df, (we use the conservative value for 40 df).
Step 4: For procedure 3, the critical t value is 2.056. For procedure 4, the critical t value is 2.021.
Step 5: Reject H_0 in favor of H_1 if $t <$ −critical t value or $t >$ critical t value. Otherwise, fail to reject H_0.
Step 6: For procedure 3, the test statistic is $t = \dfrac{\overline{x}_1 - \overline{x}_2}{\sqrt{\dfrac{s_1^2}{n_1} + \dfrac{s_2^2}{n_2}}}$. For procedure 4, the test statistic is

$$t = \frac{\overline{x}_1 - \overline{x}_2}{\sqrt{\dfrac{s_1^2(n_1 - 1) + s_2^2(n_2 - 1)}{n_1 + n_2 - 2}}\sqrt{\dfrac{1}{n_1} + \dfrac{1}{n_2}}}.$$

Variable	Procedure	Critical t value	Test statistic, t	Step 7: Make the decision.
Quality	3	2.056	1.654	Fail to reject H_0.
Reliability	4	2.021	1.050	Fail to reject H_0.
Reputation	3	2.056	1.955	Fail to reject H_0.
Value	3	2.056	−1.942	Fail to reject H_0.
Worthwhile Buy	4	2.021	−0.734	Fail to reject H_0.
Reasonable Buy	4	2.021	−8.339	Reject H_0.
Intend to Buy	4	2.021	−3.064	Reject H_0.

For each of the seven variables we construct a 95 percent confidence interval for the difference in the population means.

If we rejected H_0 in the first hypotheses tests on variance, then $\sigma_1^2 \neq \sigma_2^2$, so we use Procedure 3 to produce the 95% confidence interval, $\overline{x}_1 - \overline{x}_2 \pm t_{\alpha/2}\sqrt{\dfrac{s_1^2}{n_1} + \dfrac{s_2^2}{n_2}}$. We use a t distribution with $27 - 1 = 26$, $t_{\alpha/2} = t_{0.025} = 2.056$.
If we failed to reject H_0, then $\sigma_1^2 = \sigma_2^2$, so we use Procedure 4 to produce the 95% confidence interval, $\overline{x}_1 - \overline{x}_2 \pm t_{\alpha/2}\sqrt{\dfrac{s_1^2(n_1 - 1) + s_2^2(n_2 - 1)}{n_1 + n_2 - 2}}\sqrt{\dfrac{1}{n_1} + \dfrac{1}{n_2}}$. We use a t distribution with $27 + 27 - 2 = 52$, (we use the conservative value for 40 df) $t_{\alpha/2} = t_{0.025} = 2.021$.

Variable	Procedure	95 percent confidence interval
Quality	3	$(4.59 - 4.22) \pm (2.056)(0.224) = 0.37 \pm 0.461$ or -0.091 to 0.831.
Reliability	4	$(4.56 - 4.37) \pm (2.021)(0.181) = 0.19 \pm 0.366$ or -0.176 to 0.556.
Reputation	3	$(4.78 - 4.44) \pm (2.056)(0.174) = 0.34 \pm 0.358$ or -0.018 to 0.698.
Value	3	$(3.96 - 4.41) \pm (2.056)(0.232) = -0.45 \pm 0.477$ or -0.927 to 0.027.
Worthwhile Buy	4	$(4.00 - 4.15) \pm (2.021)(0.204) = -0.15 \pm 0.412$ or -0.562 to 0.262.
Reasonable Buy	4	$(2.51 - 4.44) \pm (2.021)(0.231) = -1.93 \pm 0.467$ or -2.397 to -1.463.
Intend to Buy	4	$(3.29 - 4.07) \pm (2.021)(0.255) = -0.78 \pm 0.515$ or -1.295 to -0.265.

34. Step 1: H_0: $\mu_1 = \mu_2$ and H_1: $\mu_1 \neq \mu_2$.
Step 2: $\alpha = 0.05$.
Step 3: Since the samples are independent and both samples are small, we checked in exercise 33 and saw that the populations have equal variances. We use a t distribution with 10 df and use procedure 4.
Step 4: The critical t values with 10 df is ± 2.228.
Step 5: Reject H_0 in favor of H_1 if $\boldsymbol{t} < -2.228$ or $\boldsymbol{t} > +2.228$. Otherwise, fail to reject H_0.
Step 6: Then

$$\boldsymbol{t} = \frac{\overline{x}_1 - \overline{x}_2}{\sqrt{\frac{s_1^2(n_1-1)+s_2^2(n_2-1)}{n_1+n_2-2}}\sqrt{\frac{1}{n_1}+\frac{1}{n_2}}} = \frac{368.86 - 353.28}{\sqrt{\frac{(1.12)^2(5)+(1.47)^2(5)}{10}}\sqrt{\frac{1}{6}+\frac{1}{6}}} = \frac{15.58}{0.7545} = 20.7.$$

Step 7: Since $\boldsymbol{t} = 20.7 > 2.228$, we reject H_0 that there is a significant difference in weight between Regular Coke and Diet Coke.

36. Step 1: H_0: $\sigma_1^2 = \sigma_2^2$ and H_1: $\sigma_1^2 \neq \sigma_2^2$.
Step 2: Use $\alpha = 0.05$.
Step 3: We use an F distribution. Since this is a two-tailed test, we use the table with 0.025 in the right tail. The df for the numerator and the denominator is $6 - 1 = 5$.
Step 4: The critical F value is 7.15.
Step 5: Reject H_0 in favor of H_1 if $\boldsymbol{F} > 7.15$. Otherwise, fail to reject H_0.
Step 6: Then $\boldsymbol{F} = \frac{(5.87)^2}{(2.79)^2} = 4.427$.
Step 7: Since $\boldsymbol{F}$ is not in the rejection region, we fail to reject H_0. The variances appear equal.

38. $s_p = \sqrt{\frac{s_1^2(n_1-1)+s_2^2(n_2-1)}{n_1+n_2-2}} = \sqrt{\frac{(5.87)^2(5)+(2.79)^2(5)}{10}} = 4.596$. The 95% confidence interval for the difference of means for Pepsi and Diet Pepsi, small independent samples with assumed equal variance (see Exercise 36): $\overline{x}_1 - \overline{x}_2 \pm t_{\alpha/2} s_p \sqrt{\frac{1}{n_1}+\frac{1}{n_2}} = (368.64 - 354.20) \pm 2.228(4.596)\sqrt{\frac{1}{6}+\frac{1}{6}}$
$= 14.44 \pm 5.91$ or 8.53 to 20.35 gm.

40. Step 1: H_0: $\mu_1 = \mu_2$ and H_1: $\mu_1 \neq \mu_2$.
Step 2: $\alpha = 0.05$.
Step 3: Since the samples are independent and both samples are small, we checked in exercise 39 and saw that the populations have equal variances. We use a t distribution with 15 df and use procedure 4.
Step 4: The critical t values with 15 df are ± 2.131.
Step 5: Reject H_0 in favor of H_1 if $\boldsymbol{t} < -2.131$ or $\boldsymbol{t} > +2.131$. Otherwise, fail to reject H_0.

Step 6: Then $t = \dfrac{\overline{x}_1 - \overline{x}_2}{\sqrt{\dfrac{s_1^2(n_1-1)+s_2^2(n_2-1)}{n_1+n_2-2}}\sqrt{\dfrac{1}{n_1}+\dfrac{1}{n_2}}} = \dfrac{15.386 - 15.666}{\sqrt{\dfrac{(0.364)^2(7)+(0.360)^2(8)}{15}}\sqrt{\dfrac{1}{8}+\dfrac{1}{9}}}$

$= \dfrac{-0.28}{0.176} = -1.591.$

Step 7: Since $t = -1.591$, we fail to reject H_0, there is not enough significant difference between the race times for 3 miles.

42. Step 1: H_0: $\sigma_1^2 = \sigma_2^2$ and H_1: $\sigma_1^2 \neq \sigma_2^2$.

Step 2: Use $\alpha = 0.05$.

Step 3: We use an F distribution. Since this is a two-tailed test, we use the table with 0.025 in the right tail. The df for the numerator 7 and the df for the denominator is 8.

Step 4: The critical F value is 4.53.

Step 5: Reject H_0 in favor of H_1 if $F > 4.53$. Otherwise, fail to reject H_0.

Step 6: Then $F = \dfrac{(0.746)^2}{(0.658)^2} = 1.285.$

Step 7: Since F is not in the rejection region, we fail to reject H_0. The variances appear equal.

We now test the hypothesis that the population means for the 5 mile times of Nebraska and Colorado are equal.

Step 1: H_0: $\mu_1 = \mu_2$ and H_1: $\mu_1 \neq \mu_2$.

Step 2: $\alpha = 0.05$.

Step 3: Since the samples are independent and both samples are small, we checked in exercise 39 and saw that the populations have equal variances. We use a t distribution with 13df and use procedure 4.

Step 4: The critical t values with 15 df is ± 2.160.

Step 5: Reject H_0 in favor of H_1 if $t < -2.160$ or $t > +2.160$. Otherwise, fail to reject H_0.

Step 6: Then $t = \dfrac{\overline{x}_1 - \overline{x}_2}{\sqrt{\dfrac{s_1^2(n_1-1)+s_2^2(n_2-1)}{n_1+n_2-2}}\sqrt{\dfrac{1}{n_1}+\dfrac{1}{n_2}}} = \dfrac{26.034 - 26.230}{\sqrt{\dfrac{(0.746)^2(7)+(0.658)^2(8)}{15}}\sqrt{\dfrac{1}{8}+\dfrac{1}{9}}}$

$= \dfrac{-0.196}{0.340} = -0.576$

Step 7: Since $t = -0.576$, we fail to reject H_0.

$s_p = \sqrt{\dfrac{s_1^2(n_1-1)+s_2^2(n_2-1)}{n_1+n_2-2}} = \sqrt{\dfrac{(0.746)^2(7)+(0.658)^2(8)}{15}} = 0.700$. The 95% confidence interval for the difference of means for the 5 mile times of Nebraska and Colorado, small independent samples with assumed equal variance (see above):

$\overline{x}_1 - \overline{x}_2 \pm t_{\alpha/2} s_p \sqrt{\dfrac{1}{n_1}+\dfrac{1}{n_2}} = (26.034 - 26.230) \pm 2.131(0.700)\sqrt{\dfrac{1}{8}+\dfrac{1}{9}} = -0.196 \pm 0.725$ or -0.921 to 0.529.

44. We now test the hypothesis that the population means are equal. For all six tests:

Step 1: H_0: $\mu_1 = \mu_2$ and H_1: $\mu_1 \neq \mu_2$.

Step 2: $\alpha = 0.05$.

Step 3: If we rejected H_0 in the previous test of variance, then $\sigma_1^2 \neq \sigma_2^2$, so we use procedure 3 and a t distribution with 3 df. If we failed to reject H_0 in the previous test of variance, then $\sigma_1^2 = \sigma_2^2$, so we use procedure 4 and a t distribution with 8 df.

Step 4: For procedure 3, the critical t value is 3.182. For procedure 4, the critical t value is 2.306.

Step 5: Reject H_0 in favor of H_1 if $t <$ −critical t value or $t >$ critical t value. Otherwise, fail to reject H_0.

Step 6: For procedure 3, the test statistic is $t = \dfrac{\overline{x}_1 - \overline{x}_2}{\sqrt{\dfrac{s_1^2}{n_1} + \dfrac{s_2^2}{n_2}}}$. For procedure 4, the test statistic is

$$t = \frac{\overline{x}_1 - \overline{x}_2}{\sqrt{\dfrac{s_1^2(n_1-1) + s_2^2(n_2-1)}{n_1+n_2-2}}\sqrt{\dfrac{1}{n_1} + \dfrac{1}{n_2}}}.$$

Variable	Procedure	Critical t value	Test statistic, t	Step 7: Make the decision.
BEF	3	3.182	−14.047	Reject H_0.
LO	4	2.306	4.564	Reject H_0.
THD	3	3.182	−1.822	Fail to reject H_0.
PF	4	2.306	3.048	Reject H_0.
Lamp	4	2.306	2.737	Reject H_0.
Blst	4	2.306	3.753	Reject H_0.

46. Step 1: H_0: $\mu_1 = \mu_2$ and H_1: $\mu_1 \neq \mu_2$.
Step 2: $\alpha = 0.01$.
Step 3: We showed in Exercise 10, Self-Testing Review 9-1 that the variance are equal, so we use procedure 4 and a t distribution with $20 + 20 - 2 = 38$ df.
Step 4: The critical t value is 2.704.
Step 5: Reject H_0 in favor of H_1 if $t < -2.704$ or if $t > 2.704$. Otherwise, fail to reject H_0.
Step 6: The test statistic is $t = \dfrac{4077 - 4275}{\sqrt{\dfrac{527^2(19) + 578^2(19)}{38}}\sqrt{\dfrac{1}{20} + \dfrac{1}{20}}} = -1.132$.
Step 7: Since $t = -1.132$, we fail to reject H_0.

48. In Exercise 11, Self-Testing Review we 9-1 reject H_0, the variance are not equal, so a 95% confidence interval is given by $\overline{x}_1 - \overline{x}_2 \pm t_{\alpha/2}\sqrt{\dfrac{s_1^2}{n_1} + \dfrac{s_2^2}{n_2}}$. We use a t distribution with 15 df. So $t_{\alpha/2} = t_{0.025} = 2.131$.
$(135.19 - 23.75) \pm (2.131)\sqrt{\dfrac{14.05^2}{16} + \dfrac{2.864^2}{16}} = 111.44 \pm 7.62$ or 103.82 to 119.06 minutes.

50. A 90% confidence interval is given by $\overline{d} \pm t_{\alpha/2}\dfrac{s_d}{\sqrt{n}}$. We use a t distribution with 15 df. So $t_{\alpha/2} = t_{0.05} = 1.753$. So $141 \pm (1.753)\dfrac{197.4}{\sqrt{16}} = 141 \pm 86.51$ or \$54.49 to \$227.51.

52. Step 1: H_0: $\mu_1 = \mu_2$ and H_1: $\mu_1 \neq \mu_2$.
Step 2: $\alpha = 0.05$.
Step 3: We showed in Exercise 15, Self-Testing Review 9-1 that the variance are equal, so we use procedure 4 and a t distribution with $11 + 32 - 2 = 41$ df. We will use the value for 40 df.
Step 4: The critical t value is 2.021.
Step 5: Reject H_0 in favor of H_1 if $t < -2.021$ or if $t > 2.021$. Otherwise, fail to reject H_0.
Step 6: The test statistic is $t = \dfrac{1077.3 - 899.8}{\sqrt{\dfrac{103.2^2(10) + 120.5^2(31)}{41}}\sqrt{\dfrac{1}{11} + \dfrac{1}{32}}} = 4.359$.
Step 7: Since $t = 4.359$, we reject H_0. The times for female and males in a 2 mile run do not have equal population means.

54. A 95% confidence interval is given by $\overline{d} \pm t_{\alpha/2}\frac{s_d}{\sqrt{n}}$. We use a t distribution with 6 df. So $t_{\alpha/2} = t_{0.025} = 2.447$. So $-0.321 \pm (2.447)\frac{0.377}{\sqrt{7}} = -0.321 \pm 0.347$ or -0.670 to 0.027.

56. A 95% confidence interval is given by $\overline{d} \pm t_{\alpha/2}\frac{s_d}{\sqrt{n}}$. We use a t distribution with 6 df. So $t_{\alpha/2} = t_{0.025} = 2.447$. So $-0.490 \pm (2.447)\frac{0.781}{\sqrt{7}} = -0.490 \pm 0.722$ or -1.212 to 0.232.

58. Step 1: H_0: $\mu_1 = \mu_2$ and H_1: $\mu_1 \neq \mu_2$.
Step 2: $\alpha = 0.01$.
Step 3: We showed in Exercise 57 Review that the variance are equal, so we use procedure 4 and a t distribution with $15 + 9 - 2 = 22$ df.
Step 4: The critical t value is 2.074.
Step 5: Reject H_0 in favor of H_1 if $t < -2.074$ or if $t > 2.074$. Otherwise, fail to reject H_0.
Step 6: The test statistic is $\boldsymbol{t} = \dfrac{10.79 - 21.36}{\sqrt{\frac{8.21^2(14) + 11.45^2(8)}{22}}\sqrt{\frac{1}{15} + \frac{1}{9}}} = -2.634$.
Step 7: Since $\boldsymbol{t} = -2.634$, we reject H_0.

60. Step 1: H_0: $\sigma_1^2 = \sigma_2^2$ and H_1: $\sigma_1^2 \neq \sigma_2^2$.
Step 2: Use $\alpha = 0.10$.
Step 3: We use an F distribution. Since this is a two-tailed test, we use the table with 0.05 in the right tail. The df for the numerator is 14 and the df for the denominator is 8. (Since there is no value for these degrees of freedom, we take the value for 12 df for the numerator and 8 df for the denominator.)
Step 4: The critical F value is 3.28.
Step 5: Reject H_0 in favor of H_1 if $\boldsymbol{F} > 3.28$. Otherwise, fail to reject H_0.
Step 6: Then $\boldsymbol{F} = \dfrac{(19.58)^2}{(18.91)^2} = 1.072$.
Step 7: Since $\boldsymbol{F}$ is not in the rejection region, we fail to reject H_0. The variances appear equal.

62. $s_p = \sqrt{\dfrac{s_1^2(n_1 - 1) + s_2^2(n_2 - 1)}{n_1 + n_2 - 2}} = \sqrt{\dfrac{(18.91)^2(8) + (19.58)^2(14)}{22}} = 19.34$. The 90% confidence interval (small independent samples with assumed equal variance) with 22 df, $t_{\alpha/2} = t_{0.05} = 1.717$
$\overline{x}_1 - \overline{x}_2 \pm t_{\alpha/2}s_p\sqrt{\frac{1}{n_1} + \frac{1}{n_2}} = (34.8 - 19.8) \pm 1.717(19.35)\sqrt{\frac{1}{9} + \frac{1}{15}} = 15 \pm 14.0$ or 1.0 to 29.0.

64. No, procedure 1 has no assumption whether the variances are equal or not.

66. A 99% confidence interval is given by $\overline{d} \pm t_{\alpha/2}\frac{s_d}{\sqrt{n}}$. We use a t distribution with 5 df. So $t_{\alpha/2} = t_{0.005} = 4.032$. So $-0.305 \pm (4.032)\frac{0.288}{\sqrt{6}} = -0.305 \pm 0.474$ or -0.779 to 0.169.
Assumption: dependent samples from normal populations.

68. We now test the hypothesis that the population means are equal. For all three test:
Step 1: H_0: $\mu_1 = \mu_2$ and H_1: $\mu_1 \neq \mu_2$.
Step 2: $\alpha = 0.05$.

Step 3: If we rejected H_0 in the previous test of variance, then $\sigma_1^2 \neq \sigma_2^2$, so we use procedure 3 and a t distribution with 3 df. If we failed to reject H_0 in the previous test of variance, then $\sigma_1^2 = \sigma_2^2$, so we use procedure 4 and a t distribution with 7 df.
Step 4: For procedure 3, the critical t value is 3.182. For procedure 4, the critical t value is 2.365.
Step 5: Reject H_0 in favor of H_1 if $t <$ −critical t value or $t >$ critical t value. Otherwise, fail to reject H_0.
Step 6: For procedure 3, the test statistic is $t = \dfrac{\overline{x}_1 - \overline{x}_2}{\sqrt{\dfrac{s_1^2}{n_1} + \dfrac{s_2^2}{n_2}}}$. For procedure 4, the test statistic is

$$t = \frac{\overline{x}_1 - \overline{x}_2}{\sqrt{\dfrac{s_1^2(n_1-1) + s_2^2(n_2-1)}{n_1+n_2-2}}\sqrt{\dfrac{1}{n_1} + \dfrac{1}{n_2}}}.$$

Variable	Procedure	Critical t value	Test statistic, t	Step 7: Make the decision.
Systolic	4	2.365	0.895	Fail to reject H_0.
Diastolic	4	2.365	2.59	Reject H_0.
Recovery	3	3.182	1.67	Fail to reject H_0.

9-3 Inference About Two Percentages

2. Step 1: H_0: $\pi_1 = \pi_2$ and H_1: $\pi_1 > \pi_2$.
Step 2: $\alpha = 0.05$.
Step 3: We use the z distribution.
Step 4: The critical z value is 1.645.
Step 5: Reject H_0 in favor of H_1 if $z > 1.645$. Otherwise, fail to reject H_0.
Step 6: Then $z = \dfrac{p_1 - p_2}{\sqrt{\dfrac{p_1(100-p_1)}{n_1} + \dfrac{p_2(100-p_2)}{n_2}}} = \dfrac{30.8108 - 19.5876}{\sqrt{27.761}} = \dfrac{11.2232}{5.2689} = 2.1301.$
Step 7: Since $z = 2.1301$ falls in the rejection region, we reject H_0. The high-risk behavior has apparently declined.

4. The confidence interval for the difference of two population percentages is: $(p_1 - p_2) \pm z_{\alpha/2}\widehat{\sigma}_{p_1-p_2}$. $-5.56 \pm (2.575)(11.749) = -5.56 \pm 30.25$ or -24.69 to 35.81.

6. Step 1: H_0: $\pi_1 = \pi_2$ and H_1: $\pi_1 \neq \pi_2$.
Step 2: $\alpha = 0.01$.
Step 3: We use the z distribution.
Step 4: The critical z values are ± 2.575.
Step 5: Reject H_0 in favor of H_1 if $z < -2.575$ or $z > +2.575$. Otherwise, fail to reject H_0.
Step 6: Then $z = \dfrac{p_1 - p_2}{\sqrt{\dfrac{p_1(100-p_1)}{n_1} + \dfrac{p_2(100-p_2)}{n_2}}} = \dfrac{-1.8873}{\sqrt{199.7174}} = -0.133.$
Step 7: Since $z = -0.133$, we fail to reject H_0 that the population percentages are equal.

8. Step 1: H_0: $\pi_1 = \pi_2$ and H_1: $\pi_1 \neq \pi_2$.
Step 2: $\alpha = 0.05$.
Step 3: We use the z distribution.

Step 4: The critical z values are ± 1.96.
Step 5: Reject H_0 in favor of H_1 if $z < -1.96$ or $z > +1.96$. Otherwise, fail to reject H_0.
Step 6: Then $z = \dfrac{p_1 - p_2}{\sqrt{\dfrac{p_1(100-p_1)}{n_1} + \dfrac{p_2(100-p_2)}{n_2}}} = \dfrac{-19.0217}{\sqrt{60.0295}} = -2.455$.
Step 7: Since $z = -2.455$ falls in the rejection region, we reject H_0.

10. Step 1: H_0: $\pi_1 = \pi_2$ and H_1: $\pi_1 > \pi_2$.
Step 2: $\alpha = 0.01$.
Step 3: We use the z distribution.
Step 4: The critical z value is 2.33.
Step 5: Reject H_0 in favor of H_1 if $z > 2.33$. Otherwise, fail to reject H_0.
Step 6: Then $z = \dfrac{p_1 - p_2}{\sqrt{\dfrac{p_1(100-p_1)}{n_1} + \dfrac{p_2(100-p_2)}{n_2}}} = \dfrac{5.5946}{\sqrt{11.4374}} = 1.654$.
Step 7: Since $z = 1.654$, we fail to reject H_0 that the population percentages are equal.

12. The confidence interval for the difference of two population percentages is: $(p_1 - p_2) \pm z_{\alpha/2}\widehat{\sigma}_{p_1-p_2}$.
$-36.4865 \pm 1.96(\sqrt{13.3318}) = -36.4865 \pm 7.1565$ or -43.6430 to -29.3300.

14. Step 1: H_0: $\pi_1 = \pi_2$ and H_1: $\pi_1 > \pi_2$.
Step 2: $\alpha = 0.05$.
Step 3: We use the z distribution.
Step 4: The critical z value is 1.645.
Step 5: Reject H_0 in favor of H_1 if $z > 1.645$. Otherwise, fail to reject H_0.
Step 6: $p_1 = 83.6634$, $p_2 = 64.0625$. Then $z = \dfrac{p_1 - p_2}{\sqrt{\dfrac{p_1(100-p_1)}{n_1} + \dfrac{p_2(100-p_2)}{n_2}}} = \dfrac{19.6009}{\sqrt{18.7571}} = 4.526$.
Step 7: Since $z = 4.526$ falls in the rejection region, we reject H_0 in favor of H_1 that the percent of smokers who are still smoking is higher in the nicotine-dependent group.

16. Step 1: H_0: $\pi_1 = \pi_2$ and H_1: $\pi_1 \neq \pi_2$.
Step 2: $\alpha = 0.10$.
Step 3: We use the z distribution.
Step 4: The critical z value is 1.645.
Step 5: Reject H_0 in favor of H_1 if $z > 1.645$ or if $z < -1.645$. Otherwise, fail to reject H_0.
Step 6: $p_1 = 93.6$, $p_2 = 88.4$. Then $z = \dfrac{p_1 - p_2}{\sqrt{\dfrac{p_1(100-p_1)}{n_1} + \dfrac{p_2(100-p_2)}{n_2}}} = \dfrac{5.2}{\sqrt{6.498}} = 2.040$.
Step 7: Since $z = 2.040$ falls in the rejection region, we reject H_0 in favor of H_1 that the percent of wafers that pass differs between the two companies.

18. Step 1: H_0: $\pi_1 = \pi_2$ and H_1: $\pi_1 > \pi_2$.
Step 2: $\alpha = 0.10$.
Step 3: We use the z distribution.
Step 4: The critical z value is 1.645.
Step 5: Reject H_0 in favor of H_1 if $z > 1.645$ of if $z < -1.645$. Otherwise, fail to reject H_0.

Step 6: $p_1 = 18.32$, $p_2 = 23.99$. Then $z = \dfrac{p_1 - p_2}{\sqrt{\dfrac{p_1(100-p_1)}{n_1} + \dfrac{p_2(100-p_2)}{n_2}}} = \dfrac{-5.67}{\sqrt{0.4062}} = -8.897.$

Step 7: Since $z = -8.897$, does not fall in the rejection region, we fail to reject H_0 in favor of H_1. Notice that the large negative value of the test which would have lead to a rejection of H_0 if we had done a left- or two-tailed test.

20. Step 1: H_0: $\pi_1 = \pi_2$ and H_1: $\pi_1 < \pi_2$.
Step 2: $\alpha = 0.01$.
Step 3: We use the z distribution.
Step 4: The critical z value is 2.33.
Step 5: Reject H_0 in favor of H_1 if $z < -2.33$. Otherwise, fail to reject H_0.
Step 6: $p_1 = 4.76$, $p_2 = 16.86$. Then $z = \dfrac{p_1 - p_2}{\sqrt{\dfrac{p_1(100-p_1)}{n_1} + \dfrac{p_2(100-p_2)}{n_2}}} = \dfrac{-12.1}{\sqrt{2.12}} = -8.310.$
Step 7: Since $z = -8.310$ falls in the rejection region, we reject H_0 in favor of H_1.

22. Step 1: H_0: $\pi_1 = \pi_2$ and H_1: $\pi_1 \neq \pi_2$.
Step 2: $\alpha = 0.10$.
Step 3: We use the z distribution.
Step 4: The critical z values are ± 1.645.
Step 5: Reject H_0 in favor of H_1 if $z < -1.645$ or $z > 1.645$. Otherwise, fail to reject H_0.
Step 6: $p_1 = 26.67$, $p_2 = 48.00$. Then $z = \dfrac{p_1 - p_2}{\sqrt{\dfrac{p_1(100-p_1)}{n_1} + \dfrac{p_2(100-p_2)}{n_2}}} = \dfrac{-21.33}{\sqrt{115.11}} = -1.988.$
Step 7: Since $z = -1.988$, we reject H_0 that the population percentages are equal.

24. Step 1: H_0: $\pi_1 = \pi_2$ and H_1: $\pi_1 \neq \pi_2$.
Step 2: $\alpha = 0.05$.
Step 3: We use the z distribution.
Step 4: The critical z values are ± 1.96.
Step 5: Reject H_0 in favor of H_1 if $z < -1.96$ or $z > 1.96$. Otherwise, fail to reject H_0.
Step 6: $p_1 = 56.44$, $p_2 = 58.04$. Then $z = \dfrac{p_1 - p_2}{\sqrt{\dfrac{p_1(100-p_1)}{n_1} + \dfrac{p_2(100-p_2)}{n_2}}} = \dfrac{-1.60}{\sqrt{41.37}} = -0.249.$
Step 7: Since $z = -0.249$ does not fall in the rejection region, we fail to reject H_0.

26. Step 1: H_0: $\pi_1 = \pi_2$ and H_1: $\pi_1 < \pi_2$.
Step 2: $\alpha = 0.10$.
Step 3: We use the z distribution.
Step 4: The critical z value is -1.28.
Step 5: Reject H_0 in favor of H_1 if $z < -1.28$. Otherwise, fail to reject H_0.
Step 6: $p_1 = 6.41$, $p_2 = 10.35$. Then $z = \dfrac{p_1 - p_2}{\sqrt{\dfrac{p_1(100-p_1)}{n_1} + \dfrac{p_2(100-p_2)}{n_2}}} = \dfrac{-3.94}{\sqrt{4.0665}} = -1.954.$
Step 7: Since $z = -1.954$, falls in the rejection region, we reject H_0 in favor of H_1.

28. The confidence interval for the difference of two population percentages is: $(p_1 - p_2) \pm z_{\alpha/2}\widehat{\sigma}_{p_1 - p_2}$.

$(91.75 - 34.37) \pm 1.96(\sqrt{8.8522}) = 57.38 \pm 5.83$ or 51.55 to 63.21.

30. The confidence interval for the difference of two population percentages is: $(p_1 - p_2) \pm z_{\alpha/2}\widehat{\sigma}_{p_1-p_2}$.
$(75.60 - 72.13) \pm 1.645(\sqrt{8.2365}) = 3.47 \pm 4.72$ or -1.25 to 8.19.

Answers to Even-Numbered Exercises

2. In Exercise 1 we fail to reject the null hypotheses that $\sigma_1^2 = \sigma_2^2$ so we use procedure 4 with a t distribution.
Step 1: H_0: $\mu_1 = \mu_2$ and H_1: $\mu_1 < \mu_2$.
Step 2: $\alpha = 0.05$.
Step 3: We use the t distribution.
Step 4: This is a one-tailed test. With $\alpha = 0.05$ and df $= n_1 + n_2 - 2 = 31$, so we use $t_{0.05} = -1.696$ (since the table does not contain 31 degrees of freedom, we take the values for 30 df and 40 df and extrapolate to find this value). Note: we chose the subscripts in Exercise 1.
Step 5: Reject H_0 in favor of H_1 if $t < -1.696$.
Step 6: Calculate the pooled standard deviation, $s_p = \sqrt{\dfrac{s_1^2(n_1-1)+s_2^2(n_2-1)}{n_1+n_2-2}}$
$= \sqrt{\dfrac{(5.67)^2(16-1)+(4.74)^2(17-1)}{16+17-2}} = \sqrt{\dfrac{841.7151}{31}} = 5.21$. The estimated standard error is $\widehat{\sigma}_{\overline{x}_1-\overline{x}_2} = s_p\sqrt{\dfrac{1}{n_1}+\dfrac{1}{n_2}} = 5.21\sqrt{\dfrac{1}{16}+\dfrac{1}{17}} = 5.21 \cdot 0.3483 = 1.815$. Computing we get
$t = \dfrac{\overline{x}_1 - \overline{x}_2}{\widehat{\sigma}_{\overline{x}_1-\overline{x}_2}} = \dfrac{13.56 - 22.11}{1.815} = -4.71$.
Step 7: Since $t < -1.696$ we reject H_0, that is, the time spent shopping by those who are touched is significantly greater than the similar time spent by those who are not touched.

4. The confidence interval for the difference of two population percentages is: $(p_1 - p_2) \pm z_{\alpha/2}\widehat{\sigma}_{p_1-p_2}$.
$(4.2 - 4.6) \pm 2.575\sqrt{0.1760} = -0.4 \pm 1.08$ or -1.48 to 0.68.

6. Since the samples are dependent, we use procedure 1 with a t test for paired differences.
Step 1: H_0: $\mu_d = 0$ and H_1: $\mu_d < 0$.
Step 2: $\alpha = 0.05$.
Step 3: We use the t distribution with $n = 8$.
Step 4: This is a one-tail test. With $\alpha = 0.05$ and df $= 7$, use $t_{0.05} = -1.895$.
Step 5: Reject H_0 in favor of H_1 if $t < -1.895$.
Step 6: Using MINITAB to calculate $\overline{d}$ and s_d:

Twin A	Twin B	d_i	$d_i - \overline{d}$	$(d_i - \overline{d})^2$
34	29	5	9.75	95.062
18	42	−24	−19.25	370.562
39	33	6	10.75	115.562
31	40	−9	−4.25	18.062
28	38	−10	−5.25	27.562
26	40	−14	−9.25	85.563
28	27	1	5.75	33.063
22	15	7	11.75	138.063
		−38		883.499

$\overline{d} = \frac{-38}{8} = -4.75$ and $s_d = \sqrt{\frac{883.499}{7}} = \sqrt{126.2141} = 11.2345$. Using $t = \frac{\overline{d} - \mu_d}{s_d/\sqrt{n}} = \frac{-4.75 - 0}{11.2345/\sqrt{8}} = -1.1959$.

Step 7: Since the t is not in the rejection region, we fail to reject H_0 that there's no significant difference in the length of time each twin remained with the puzzle.

8. Sample sizes are > 30 so use procedure 2 with a z distribution.
Step 1: H_0: $\mu_1 = \mu_2$ and H_1: $\mu_1 \neq \mu_2$.
Step 2: $\alpha = 0.05$.
Step 3: z distribution.
Step 4: This is a two-tailed test with $\alpha = 0.05$, thus the critical z values are ± 1.96.
Step 5: Reject H_0 in favor of H_1 if $z < -1.96$ or $z > 1.96$. Otherwise, fail to reject H_0.
Step 6: The standard error of the difference between means $\widehat{\sigma}_{\overline{x}_1 - \overline{x}_2} = \sqrt{\frac{s_1^2}{n_1} + \frac{s_2^2}{n_2}} = \sqrt{\frac{2.9^2}{141} + \frac{3.2^2}{123}}$ $= \sqrt{0.14290} = 0.3780$. Then $z = \frac{\overline{x}_1 - \overline{x}_2}{\sqrt{\frac{s_1^2}{n_1} + \frac{s_2^2}{n_2}}} = \frac{13.9 - 13.5}{0.3780} = \frac{0.4}{0.3780} = 1.058$.
Step 7: Since $z = 1.058$ does not fall in the rejection region, we fail to reject H_0 that there is no difference in the education of the population of caregivers who fell into these two groups.

10. Step 1: H_0: $\sigma_1^2 = \sigma_2^2$ and H_1: $\sigma_1^2 \neq \sigma_2^2$.
Step 2: $\alpha = 0.05$.
Step 3: Because this is a test of variance we use an F distribution.
Step 4: Since $\alpha = 0.05$ and this is a two-tailed test, we use the table for $\alpha = 0.025$. Since there are 16 members in each group we use 15 degrees of freedom for both the numerator and denominator. The critical F value is 2.86.
Step 5: Reject H_0 in favor of H_1 if $F > 2.86$. Otherwise, fail to reject H_0.
Step 6: Then $F = \frac{s_1^2}{s_2^2} = \frac{(0.99)^2}{(0.81)^2} = \frac{0.9801}{0.6561} = 1.494$.
Step 7: Since $F = 1.494 < 2.86$, we fail to reject H_0 that the population variances are equal.

12. The formula for a conference interval for $\mu_1 - \mu_2$ is: $\overline{x}_1 - \overline{x}_2 \pm t_{\alpha/2} s_p \sqrt{\frac{1}{n_1} + \frac{1}{n_2}}$. With $\alpha = 0.01$ and $\text{df} = n_1 + n_2 - 2 = 30$, so we use $t_{\alpha/2} = t_{0.005} = 2.750$. Then the conference interval for $\mu_1 - \mu_2$ is:
$(25.80 - 25.58) \pm (2.750)\sqrt{\frac{(0.99)^2(16-1) + (0.81)^2(16-1)}{16+16-2}}\sqrt{\frac{1}{16} + \frac{1}{16}} = 0.22 \pm 0.879$ or -0.659 to 1.099.

14. Since the sample sizes are large than 30, we use procedure 2 with a z distribution.
Step 1: H_0: $\mu_1 = \mu_2$ and H_1: $\mu_1 \neq \mu_2$.
Step 2: $\alpha = 0.01$.
Step 3: z distribution.
Step 4: This is a two-tailed test with $\alpha = 0.01$, thus the critical z values are ± 2.575.
Step 5: Reject H_0 in favor of H_1 if $z < -2.575$ or if $z > 2.575$. Otherwise, fail to reject H_0.
Step 6: $\widehat{\sigma}_{\overline{x}_1 - \overline{x}_2} = \sqrt{\frac{\sigma_1^2}{n_1} + \frac{\sigma_2^2}{n_2}} = \sqrt{\frac{(13.28)^2}{72} + \frac{(13.52)^2}{55}} = \sqrt{5.7729} = 2.4027$. Then
$z = \frac{\overline{x}_1 - \overline{x}_2}{\widehat{\sigma}_{\overline{x}_1 - \overline{x}_2}} = \frac{77.04 - 69.33}{2.4027} = \frac{7.71}{2.4027} = 3.2089$.

Step 7: Since $z = 3.2089$ falls in the rejection region, we reject H_0. The attitudes of the teachers are different in the two groups.

16. To obtain a 99% confidence interval for $\mu_1 - \mu_2$, we need $z_{\alpha/2} = z_{0.005} = 2.575$. Then we compute:

$$\overline{x}_1 - \overline{x}_2 \pm z_{\alpha/2}\sqrt{\frac{\sigma_1^2}{n_1} + \frac{\sigma_2^2}{n_2}} = (42.0 - 37.2) \pm (2.575)\sqrt{\frac{(7.65)^2}{222} + \frac{(6.83)^2}{64}}$$
$$= 0.83 \pm (2.575)(0.9962) = 0.83 \pm 2.57 \text{ or } -1.74 \text{ to } 3.40.$$

18. Since the sample sizes are larger than 30 we use procedure 2 with a z distribution.
Step 1: H_0: $\mu_1 = \mu_2$ and H_1: $\mu_1 \neq \mu_2$.
Step 2: $\alpha = 0.01$.
Step 3: z distribution.
Step 4: This is a two-tailed test with $\alpha = 0.01$, thus the critical z values are ± 2.575.
Step 5: Reject H_0 in favor of H_1 if $z < -2.575$ or if $z > 2.575$. Otherwise, fail to reject H_0.

Step 6: $\hat{\sigma}_{\overline{x}_1 - \overline{x}_2} = \sqrt{\frac{\sigma_1^2}{n_1} + \frac{\sigma_2^2}{n_2}} = \sqrt{\frac{(0.83)^2}{84} + \frac{(0.97)^2}{32}} = \sqrt{0.03760} = 0.1939.$

Then $z = \frac{\overline{x}_1 - \overline{x}_2}{\hat{\sigma}_{\overline{x}_1 - \overline{x}_2}} = \frac{4.02 - 2.93}{0.1939} = 5.621.$

Step 7: Since $z = 5.621$ falls in the rejection region, we reject H_0, there is a significant difference of opinion about testing and test uses between pre-service and in-service teachers.

20. With $\alpha = 0.01$ and df $= 7$, use $t_{\alpha/2} = t_{0.005} = 3.499$. We next calculate $\overline{d}$ and s_d.

State		Medium Repslotion Time	d_i	$d_i - \overline{d}$	$(d_i - \overline{d})^2$
Alabama	55,977	37,170	18,807	−2888.5	8,343,432.25
California	72,642	45,196	27,446	5740.5	33,068,250.25
Delaware	78,371	45,025	33,346	11,650.5	135,734,150.25
Idaho	51,615	38,551	13,064	−8,631.5	74,502,792.25
Mississippi	54,843	38,729	16,223	−5472.5	29,948,256.25
New Hampshire	62,678	38,729	23,949	2,253.5	5,078,262.25
Virginia	62,785	39,227	23,558	1,862.5	3,468,906.25
Wyoming	56,574	39,403	17,171	−4,524.5	20,471,100.25
			173,564		310,615,150.0

So $\overline{d} = \frac{173,564}{8} = 21695.5$ and $s_d = \sqrt{\frac{310,615,150}{7}} = 6,661.4$. Then

$\overline{d} \pm t_{\alpha/2}\frac{s_d}{\sqrt{n}} = 21695.5 \pm 3.499\frac{6,661.4}{\sqrt{8}} = 21695.5 \pm 8240.7$ or \$13,454.80 to \$29,936.20. (MINITAB will report this as \$13,453 to \$29,938.)
The only assumption is that the difference is normally distributed.

22. Since the population variances are not equal, we use procedure 3 and a t distribution.
Step 1: H_0: $\mu_1 = \mu_2$ and H_1: $\mu_1 < \mu_2$.
Step 2: $\alpha = 0.01$.
Step 3: t distribution.
Step 4: This is a one-tail test with $\alpha = 0.01$. The degree of freedom is the smaller of $n_1 - 1$ or $n_2 - 1$, $12 - 1$ or $10 - 1$, which is 9. The critical value is $t_{0.01} = -2.821$.
Step 5: Reject H_0 in favor of H_1 if $t < -2.821$. Otherwise, fail to reject H_0.

Step 6: Then $t = \frac{\overline{x}_1 - \overline{x}_2}{\sqrt{\frac{s_1^2}{n_1} + \frac{s_2^2}{n_2}}} = \frac{1 - 6}{\sqrt{\frac{(0.6)^2}{12} + \frac{(3)^2}{10}}} = \frac{-5}{\sqrt{0.93}} = -5.185.$

Step 7: Since $t = -5.185$ falls in the rejection region, we reject H_0. Alcoholics have a higher mean rating on the depression scale than do normal subjects.

24. In Exercise 23 we rejected the null hypotheses that $\sigma_1^2 = \sigma_2^2$ so we use procedure 3 with a t distribution. Also in Exercise 23, we chose the subscript 1 to represent the SACH group and subscript 2 to represent the CC II group.
Step 1: H_0: $\mu_1 = \mu_2$ and H_1: $\mu_1 \neq \mu_2$.
Step 2: $\alpha = 0.01$.
Step 3: t distribution.
Step 4: This is a two-tailed test with $\alpha = 0.01$. The degree of freedom is 9 since both groups have 10 members. The critical values are $t_{0.005} = \pm 3.250$.
Step 5: Reject H_0 in favor of H_1 if $t < -3.250$ or if $t > 3.250$. Otherwise, fail to reject H_0.
Step 6: Then

$$t = \frac{\overline{x}_1 - \overline{x}_2}{\sqrt{\frac{s_1^2}{n_1} + \frac{s_2^2}{n_2}}} = \frac{0.764 - 0.766}{\sqrt{\frac{(0.0011)^2}{10} + \frac{(0.00054)^2}{10}}} = \frac{-0.002}{\sqrt{0.0000001501}} = \frac{-0.002}{0.0003875} = -5.161.$$

Step 7: Since $t = -5.161$ falls in the rejection region, we reject H_0, there is a difference in the population mean step length between the two types of prosthetic feet.

26. To obtain a 95% confidence interval for $\pi_1 - \pi_2$, we need $z_{\alpha/2} = z_{0.025} = 1.96$. Then we compute:

$$(p_1 - p_2) \pm z_{\alpha/2}\sqrt{\frac{\sigma_1^2}{n_1} + \frac{\sigma_2^2}{n_2}} = (47.92 - 62.32) \pm (1.96)\sqrt{\frac{47.92 \cdot 52.08}{96} + \frac{62.32 \cdot 37.68}{69}}$$
$$= -14.4 \pm (1.96)(7.747) = -14.4 \pm 15.2 \text{ or } -29.6 \text{ to } 0.8.$$

28. For males $np = 100 < 500$, sample too small.
Author's note: This exercise was constructed using real data, which does not always produce "clean" results. In this case, the expected value is less than five hundred. To have a more valid study, a larger sample is necessary.

30. To obtain a 99% confidence interval for μ_d is completed using: $\overline{d} \pm t_{\alpha/2}\frac{s_d}{\sqrt{n}}$. With $\alpha = 0.01$ and df $= 4$, use $t_{\alpha/2} = t_{.005} = 4.604$. Then we compute:

$$\overline{d} \pm t_{\alpha/2}\frac{s_d}{\sqrt{n}} = (-0.986) \pm (4.604)\frac{0.5761}{\sqrt{5}}$$
$$= -0.986 \pm (4.604)(0.258) = -0.986 \pm 1.188 \text{ or } -2.174 \text{ to } 0.202.$$

32. Since the sample sizes are larger than 30 we use procedure 2 with a z distribution.
Step 1: H_0: $\mu_1 = \mu_2$ and H_1: $\mu_1 \neq \mu_2$.
Step 2: $\alpha = 0.05$.
Step 3: z distribution.
Step 4: This is a two-tailed test with $\alpha = 0.05$, thus the critical z values are ± 1.96.
Step 5: Reject H_0 in favor of H_1 if $z < -1.96$ or if $z > 1.96$. Otherwise, fail to reject H_0.
Step 6: $\widehat{\sigma}_{\overline{x}_1 - \overline{x}_2} = \sqrt{\frac{\sigma_1^2}{n_1} + \frac{\sigma_2^2}{n_2}} = \sqrt{\frac{16.1^2}{269} + \frac{11.4^2}{253}} = \sqrt{1.4773} = 1.2154$. Then

$$z = \frac{\overline{x}_1 - \overline{x}_2}{\widehat{\sigma}_{\overline{x}_1 - \overline{x}_2}} = \frac{90.57 - 96.51}{1.2154} = \frac{-5.95}{1.2154} = -4.887.$$

Step 7: Since $z = 4.887$ falls in the rejection region, we reject H_0, that is, there is a significant difference between the population mean IQ scores for the two groups.

34. In Exercise 33 we failed to reject the null hypotheses that $\sigma_1^2 = \sigma_2^2$ so in this exercise we use procedure 4 with a t distribution.

Step 1: H_0: $\mu_1 = \mu_2$ and H_1: $\mu_1 > \mu_2$.

Step 2: $\alpha = 0.05$.

Step 3: We use the t distribution.

Step 4: This is a one-tailed test. With $\alpha = 0.05$ and 40 df we use $t_{0.05} = 1.684$.

Step 5: Reject H_0 in favor of H_1 if $t > 1.684$. Otherwise, fail to reject H_0.

Step 6: Calculate the pooled standard deviation, $s_p = \sqrt{\dfrac{s_1^2(n_1-1)+s_2^2(n_2-1)}{n_1+n_2-2}}$ $= \sqrt{\dfrac{(0.792)^2(21-1)+(0.689)^2(21-1)}{21+21-2}} = \sqrt{0.5510} = 0.7423$. The estimated standard error is $\widehat{\sigma}_{\overline{x}_1-\overline{x}_2} = s_p\sqrt{\dfrac{1}{n_1}+\dfrac{1}{n_2}} = 0.7423 \cdot \sqrt{\dfrac{1}{21}+\dfrac{1}{21}} = 0.7423 \cdot 0.3086 = 0.2291$. Thus $t = \dfrac{\overline{x}_1-\overline{x}_2}{\widehat{\sigma}_{\overline{x}_1-\overline{x}_2}} = \dfrac{2.685-2.480}{0.2291} = 0.895$.

Step 7: Since $t = 0.895$ does not fall within the rejection region, we fail to reject H_0, that is there is not a significant difference between the mean GPA of the students who stay at home and those who go away.

36. Since the samples are independent and the sample sizes are large so we conduct a two sample test on percentages with a z distribution.

Step 1: H_0: $\pi_1 = \pi_2$ and H_1: $\pi_1 < \pi_2$.

Step 2: $\alpha = 0.01$.

Step 3: We use the z distribution.

Step 4: This is a one-tailed test, with $\alpha = 0.01$ the critical z value is -2.33.

Step 5: Reject H_0 in favor of H_1 if $z < -2.33$. Otherwise, fail to reject H_0.

Step 6: Calculate the percentages: $p_1 = \dfrac{17}{60} = 28.333\%$ and $p_2 = \dfrac{32}{58} = 55.172\%$.
Calculate the standard error of the difference between percentages:

$$\widehat{\sigma}_{p_1-p_2} = \sqrt{\frac{p_1(100-p_1)}{n_1}+\frac{p_2(100-p_2)}{n_2}} = \sqrt{\frac{28.333\cdot 71.667}{60}+\frac{55.172\cdot 44.828}{58}}$$

$= \sqrt{76.4846} = 8.7455$ Thus $z = \dfrac{p_1-p_2}{\widehat{\sigma}_{p_1-p_2}} = \dfrac{28.333-55.172}{8.7455} = \dfrac{-26.839}{8.7455} = -3.069$.

Step 7: Since $z = -3.069$ falls in the rejection region, we reject H_0, thus the percentage of patients who use NORVASC and experience improvement is significantly greater than those who were given a placebo.

38. In Exercise 37 we rejected the null hypotheses that $\sigma_1^2 = \sigma_2^2$ so we use procedure 3 with a t distribution.

Step 1: H_0: $\mu_1 = \mu_2$ and H_1: $\mu_1 \neq \mu_2$.

Step 2: $\alpha = 0.10$.

Step 3: t distribution.

Step 4: This is a two-tailed test with $\alpha = 0.10$. The degree of freedom is 6 since both groups have 7 members. The critical values are $t_{0.05} = \pm 1.943$.

Step 5: Reject H_0 in favor of H_1 if $t < -1.943$ or if $t > 1.943$. Otherwise, fail to reject H_0.

Step 6: Then $t = \dfrac{\overline{x}_1-\overline{x}_2}{\sqrt{\dfrac{s_1^2}{n_1}+\dfrac{s_2^2}{n_2}}} = \dfrac{0.7143-0.7286}{\sqrt{\dfrac{(0.1464)^2}{7}+\dfrac{(0.0488)^2}{7}}} = \dfrac{-0.0143}{\sqrt{0.003405}} = \dfrac{-0.0143}{0.05832} = -0.245$.

Step 7: Since $t = -0.245$ does not fall within the rejection region, we fail to reject H_0.

40. Step 1: H_0: $\sigma_1^2 = \sigma_2^2$ and H_1: $\sigma_1^2 \neq \sigma_2^2$.

Step 2: $\alpha = 0.05$.

Step 3: Because this is a test of variance we use an F distribution.

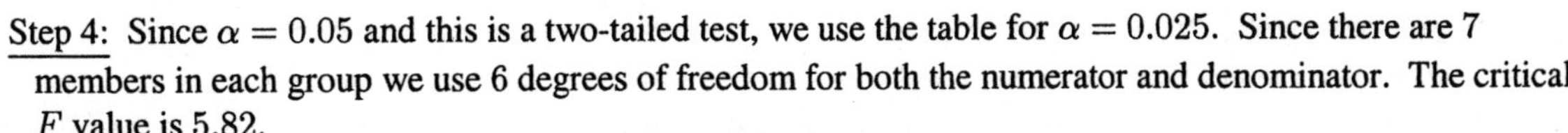

Step 4: Since $\alpha = 0.05$ and this is a two-tailed test, we use the table for $\alpha = 0.025$. Since there are 7 members in each group we use 6 degrees of freedom for both the numerator and denominator. The critical F value is 5.82.

Step 5: Reject H_0 in favor of H_1 if $\boldsymbol{F} > 5.82$. Otherwise, fail to reject H_0.

Step 6: Then $\boldsymbol{F} = \dfrac{s_1^2}{s_2^2} = \dfrac{(0.1113)^2}{(0.0535)^2} = \dfrac{0.012388}{0.002862} = 4.33.$

Step 7: Since $\boldsymbol{F} = 4.33 < 5.82$, we fail to reject H_0 that the population variances are equal for cadmium.

42. Step 1: H_0: $\sigma_1^2 = \sigma_2^2$ and H_1: $\sigma_1^2 \neq \sigma_2^2$.

Step 2: $\alpha = 0.02$.

Step 3: Because this is a test of variance we use an F distribution.

Step 4: Since $\alpha = 0.02$ and this is a two-tailed test, we use the table for $\alpha = 0.01$. Since there are 7 members in each group we use 6 degrees of freedom for both the numerator and denominator. The critical F value is 8.47.

Step 5: Reject H_0 in favor of H_1 if $\boldsymbol{F} > 8.47$. Otherwise, fail to reject H_0.

Step 6: Then $\boldsymbol{F} = \dfrac{s_1^2}{s_2^2} = \dfrac{(1.676)^2}{(0.308)^2} = \dfrac{2.80898}{0.09486} = 29.6.$

Step 7: Since $\boldsymbol{F} = 29.6 > 8.47$, we reject H_0 that the population variances are not equal for lead.

44. Step 1: $\sigma_1^2 = \sigma_2^2$ and H_1: $\sigma_1^2 > \sigma_2^2$.

Step 2: Use $\alpha = 0.01$.

Step 3: We use an F distribution. Since this is a right-tailed test, we use the table with 0.01 in the right tail.

Step 4: We use the table for $\alpha = 0.01$. Since there are 7 members in each group we use 6 degrees of freedom for both the numerator and denominator. The critical F value is 8.47.

Step 5: Reject H_0 in favor of H_1 if $\boldsymbol{F} > 8.47$. Otherwise, fail to reject H_0.

Step 6: The $\boldsymbol{F} = \dfrac{(5.32)^2}{(1.915)^2} = 7.718.$

Step 7: Since $\boldsymbol{F} = 7.718$, we fall to reject H_0, there is not enough evidence to show that the population variance of copper for lab 4 is greater than for lab 3.

46. In Exercise 44 we rejected the null hypotheses that $\sigma_1^2 = \sigma_2^2$ so we use procedure 3 with a t distribution.

Step 1: H_0: $\mu_1 = \mu_2$ and H_1: $\mu_1 \neq \mu_2$.

Step 2: $\alpha = 0.10$.

Step 3: t distribution.

Step 4: This is a two-tailed test with $\alpha = 0.10$. The degree of freedom is 9 since both groups have 10 members. The critical values are $t_{0.05} = \pm 1.833$.

Step 5: Reject H_0 in favor of H_1 if $\boldsymbol{t} < -1.833$ or if $\boldsymbol{t} > 1.833$. Otherwise, fail to reject H_0.

Step 6: Then $\boldsymbol{t} = \dfrac{\overline{x}_1 - \overline{x}_2}{\sqrt{\dfrac{s_1^2}{n_1} + \dfrac{s_2^2}{n_2}}} = \dfrac{6.461 - 7.068}{\sqrt{\dfrac{(1.602)^2}{10} + \dfrac{(0.551)^2}{10}}} = \dfrac{-0.607}{\sqrt{0.28700}} = \dfrac{-0.607}{0.53572} = 1.133.$

Step 7: Since $\boldsymbol{t} = 1.133$, we fall to reject H_0, the population means for pH are equal.

48. To obtain a 99% confidence interval for $\pi_1 - \pi_2$, we need $z_{\alpha/2} = z_{0.005} = 2.575$. Then we compute:

$$(p_1 - p_2) \pm z_{\alpha/2}\sqrt{\frac{\sigma_1^2}{n_1} + \frac{\sigma_2^2}{n_2}} = (29.34 - 15.38) \pm (2.575)\sqrt{\frac{29.34 \cdot 70.66}{167} + \frac{15.38 \cdot 84.62}{143}}$$
$$= 13.96 \pm (2.575)(4.638) = 13.96 \pm 11.94 \text{ or } 2.02 \text{ to } 25.9.$$

50. In Exercise 49 we failed to reject the null hypotheses that $\sigma_1^2 = \sigma_2^2$ so in this exercise we use procedure 4 with a t distribution.

Step 1: H_0: $\mu_1 = \mu_2$ and H_1: $\mu_1 \neq \mu_2$.

Step 2: $\alpha = 0.05$.
Step 3: We use the t distribution.
Step 4: This is a two-tailed test. With $\alpha = 0.05$ and 48 df we use $t_{\alpha/2} = t_{0.025} = 2.010$ (we took the average of the 40 df and the 60 df values).
Step 5: Reject H_0 in favor of H_1 if $t < -2.010$ or if $t > 2.010$. Otherwise, fail to reject H_0.
Step 6: Calculate the pooled standard deviation, $s_p = \sqrt{\dfrac{s_1^2(n_1-1)+s_2^2(n_2-1)}{n_1+n_2-2}}$ $= \sqrt{\dfrac{(6.32)^2(20-1)+(5.88)^2(30-1)}{20+30-2}} = \sqrt{36.6992} = 6.058$. The estimated standard error is $\widehat{\sigma}_{\overline{x}_1-\overline{x}_2} = s_p\sqrt{\dfrac{1}{n_1}+\dfrac{1}{n_2}} = 6.058 \cdot \sqrt{\dfrac{1}{20}+\dfrac{1}{30}} = 6.058 \cdot 0.289 = 1.751$. Thus $t = \dfrac{\overline{x}_1-\overline{x}_2}{\widehat{\sigma}_{\overline{x}_1-\overline{x}_2}} = \dfrac{26.28-27.56}{1.751} = \dfrac{-1.28}{1.751} = -0.731$.
Step 7: Since $t = -0.731$, we fail to reject H_0, that is there is not a significant difference in the length of the population means.

52. For all five test:
Step 1: H_0: $\pi_1 = \pi_2$ and H_1: $\pi_1 \neq \pi_2$.
Step 2: $\alpha = 0.05$.
Step 3: We use the z distribution.
Step 4: This is a two-tailed test, with $\alpha = 0.05$ the critical z value is ±1.96.
Step 5: Reject H_0 in favor of H_1 if $z < -1.96$ or if $z > 1.96$. Otherwise, fail to reject H_0.
Step 6: Calculate the percentages, the standard error of the difference between percentages:
$\widehat{\sigma}_{p_1-p_2} = \sqrt{\dfrac{p_1(100-p_1)}{n_1}+\dfrac{p_2(100-p_2)}{n_2}}$ and $z = \dfrac{p_1-p_2}{\widehat{\sigma}_{p_1-p_2}}$.

Question	p_1	p_2	$z = \dfrac{p_1-p_2}{\widehat{\sigma}_{p_1-p_2}}$	Step 7: Make the decision.
1	40	50	$z = \dfrac{-10}{9.8995} = -1.010$	Fail to reject H_0.
2	78	64	$z = \dfrac{14}{8.9666} = 1.561$	Fail to reject H_0.
3	64	64	$z = \dfrac{0}{9.6000} = 0$	Fail to reject H_0.
4	32	46	$z = \dfrac{-14}{9.6540} = -1.450$	Fail to reject H_0.
5	56	62	$z = \dfrac{-6}{9.8184} = -0.611$	Fail to reject H_0.

54. For all five tests we have two dependent samples taken from normally distributed populations, so we conduct a paired t test.
Step 1: H_0: $\mu_d = 0$ and H_1: $\mu_d \neq 0$.
Step 2: $\alpha = 0.05$.
Step 3: We used a paired t test, with procedure 1 and a t distribution with $12 - 1 = 11$ df.
Step 4: The critical t value is 2.201.
Step 5: Reject H_0 in favor of H_1 if $t < -2.201$ or $t > 2.201$. Otherwise, fail to reject H_0.
Step 6: For procedure 1 the test statistic is $t = \dfrac{\overline{d}-\mu_d}{s_d/\sqrt{n}}$.

Variable	Test statistic, t	Step 7: Make the decision.
SBP	$\dfrac{20.33}{24.28/\sqrt{12}} = 2.90$	Reject H_0.
DBP	$\dfrac{8.33}{15.65/\sqrt{12}} = 1.84$	Fail to reject H_0.
SNa	$\dfrac{2.417}{1.832/\sqrt{12}} = 4.57$	Reject H_0.
SK	$\dfrac{-1.142}{0.789/\sqrt{12}} = -5.01$	Reject H_0.
PAC	$\dfrac{1035}{937/\sqrt{12}} = 3.83$	Reject H_0.

56. Step 1: H_0: $\pi_1 = \pi_2$ and H_1: $\pi_1 \neq \pi_2$.
Step 2: $\alpha = 0.10$.
Step 3: We use the z distribution.
Step 4: This is a two-tailed test, with $\alpha = 0.10$ the critical z value is ± 1.645.
Step 5: Reject H_0 in favor of H_1 if $z < -1.645$ or if $z > 1.645$. Otherwise, fail to reject H_0.
Step 6: Calculate the percentages: $p_1 = \dfrac{8}{26} = 30.77\%$ and $p_2 = \dfrac{7}{12} = 58.33\%$.
Calculate the standard error of the difference between percentages:

$$\hat{\sigma}_{p_1-p_2} = \sqrt{\frac{p_1(100-p_1)}{n_1} + \frac{p_2(100-p_2)}{n_2}} = \sqrt{\frac{30.77 \cdot 69.23}{26} + \frac{58.33 \cdot 41.67}{12}} = \sqrt{284.476} = 16.866.$$

$$\text{Thus } z = \frac{p_1 - p_2}{\hat{\sigma}_{p_1-p_2}} = \frac{30.77 - 58.33}{16.866} = \frac{-27.56}{16.866} = -1.634.$$

Step 7: Since $z = -1.634$, we fail to reject H_0.

58. Step 1: H_0: $\pi_1 = \pi_2$ and H_1: $\pi_1 \neq \pi_2$.
Step 2: $\alpha = 0.05$.
Step 3: We use the z distribution.
Step 4: This is a two-tailed test, with $\alpha = 0.05$ the critical z value is ± 1.96.
Step 5: Reject H_0 in favor of H_1 if $z < -1.96$ or if $z > 1.96$. Otherwise, fail to reject H_0.
Step 6: Calculate the percentages: $p_1 = \dfrac{12}{26} = 46.15\%$ and $p_2 = \dfrac{6}{12} = 50\%$.
Calculate the standard error of the difference between percentages:

$$\hat{\sigma}_{p_1-p_2} = \sqrt{\frac{p_1(100-p_1)}{n_1} + \frac{p_2(100-p_2)}{n_2}} = \sqrt{\frac{46.15 \cdot 53.85}{26} + \frac{50 \cdot 50}{12}} = \sqrt{303.918} = 17.433. \text{ Thus}$$

$$z = \frac{p_1 - p_2}{\hat{\sigma}_{p_1-p_2}} = \frac{46.15 - 50}{17.433} = \frac{-3.85}{17.433} = -0.221.$$

Step 7: Since $z = -0.221$, we fail to reject H_0.

60. Step 1: H_0: $\mu_d = 0$ and H_1: $\mu_d > 0$.
Step 2: $\alpha = 0.01$.
Step 3: We use a paired t test and a t distribution with 3 df
Step 4: The critical value of $t = 4.541$.
Step 5: Reject H_0 in favor of H_1 if $t > 4.541$. Otherwise, fail to reject H_0.

Step 6: Then $t = \dfrac{\overline{d} - \mu_d}{\dfrac{s_d}{\sqrt{n}}} = \dfrac{27.8}{24.1/\sqrt{4}} = 2.307.$

Step 7: Since $t = 2.307$ does not fall in the rejection region, we fail to reject H_0.

62. Step 1: H_0: $\mu_d = 0$ and H_1: $\mu_d > 0$.
Step 2: $\alpha = 0.05$.
Step 3: We use a paired t test and a t distribution with 7 df
Step 4: The critical value of $t = 1.895$.
Step 5: Reject H_0 in favor of H_1 if $t > 1.895$. Otherwise, fail to reject H_0.
Step 6: Then $t = \dfrac{\overline{d} - \mu_d}{\dfrac{s_d}{\sqrt{n}}} = \dfrac{0.086}{0.499/\sqrt{8}} = 0.487.$
Step 7: Since $t = 0.487$ does not fall in the rejection region, we fail to reject H_0.

64.

Pretest	27.79	25.59	21.53	25.25	17.49	25.14	19.59	21.40
Posttest	28.01	25.42	21.67	26.03	17.91	25.25	20.19	21.67
Changes	0.22	−0.17	0.14	0.78	0.42	0.11	0.60	0.27

66. In Exercise 65 we failed to reject the null hypotheses that $\sigma_1^2 = \sigma_2^2$ so in this exercise we use procedure 4 with a t distribution.
Step 1: H_0: $\mu_1 = \mu_2$ and H_1: $\mu_1 < \mu_2$.
Step 2: $\alpha = 0.05$.
Step 3: We use the t distribution.
Step 4: This is a one-tailed test. With $\alpha = 0.05$ and 14 df we use $t_\alpha = t_{0.05} = 1.761$.
Step 5: Reject H_0 in favor of H_1 if $t < -1.761$. Otherwise, fail to reject H_0.
Step 6: Calculate the pooled standard deviation, $s_p = \sqrt{\dfrac{s_1^2(n_1 - 1) + s_2^2(n_2 - 1)}{n_1 + n_2 - 2}}$
$= \sqrt{\dfrac{(0.499)^2(8-1) + (0.299)^2(8-1)}{8+8-2}} = \sqrt{0.169201} = 0.411$. The estimated standard error is $\widehat{\sigma}_{\overline{x}_1 - \overline{x}_2} = s_p\sqrt{\dfrac{1}{n_1} + \dfrac{1}{n_2}} = 0.5817 \cdot \sqrt{\dfrac{1}{8} + \dfrac{1}{8}} = 0.411 \cdot 0.5 = 0.2055$. Thus
$t = \dfrac{\overline{x}_1 - \overline{x}_2}{\widehat{\sigma}_{\overline{x}_1 - \overline{x}_2}} = \dfrac{-0.086 - 0.296}{0.2055} = \dfrac{-0.382}{0.2055} = -1.86.$
Step 7: Since $t = -1.86$, we reject H_0, the body fat percentage change is greater for the experimental group.

Chapter 10 Analysis of Variance

Answers to Even-Numbered Self-Testing Review

10-1 Analysis of Variance: Purpose and Procedure

2. The null hypothesis states that all the population means are equal. The alternative hypothesis says at least one population mean is different.

4. The grand mean, $\overline{\overline{X}}$, is the mean of all the values from all the samples or factor groups. It can be computed by adding all these values and dividing by the total number of items. Alternatively, if the means of each of the samples are known, then $\overline{\overline{X}}$ can be computed by a weighted average of these means. That is, you can multiply each sample mean by the number of items in that sample, add all these products together, and then divide by T (the total number of items in all the samples).

6. The population variation between factors should always be greater than or equal to the variation within the population. In an ANOVA test, any time the test ratio produces a number less than one it is due to sampling error.

8. The $\text{df}_{\text{num}} = 6 - 1 = 5$, and the $\text{df}_{\text{den}} = 35 - 6 = 29$. So $\boldsymbol{F} = 3.73$.

10. The $\text{df}_{\text{num}} = 3 - 1 = 2$, and the $\text{df}_{\text{den}} = 29 - 3 = 26$. So $\boldsymbol{F} = 5.53$.

12. $\boldsymbol{F} = \dfrac{215.23}{73.81} = 2.9160.$

14. False. It's almost certain that these two estimated values will differ even when we fail to reject H_0 as being true.

10-2 An ANOVA Example

2. Step 1: H_0: All population means are equal, and H_1: Not all population means are equal.
   Step 2: $\alpha = 0.05$.
   Step 3: We use an F distribution with $\text{df}_{\text{num}} = 3 - 1 = 2$, and $\text{df}_{\text{den}} = 19 - 3 = 16$.
   Step 4: The critical F value is $F_{2,\,16,\,\alpha=0.05} = 3.63$.
   Step 5: Reject H_0 in favor of H_1 if $\boldsymbol{F} > 3.63$. Otherwise, fail to reject H_0.
   Step 6: The sample means are 13, 9.83, and 4.375. The d^2 values are: 106.00, 224.83, and 55.875. So
   $\widehat{\sigma}^2_{\text{within}} = \dfrac{106 + 224.83 + 55.875}{19 - 3} = \dfrac{386.705}{16} = 24.169$. Then
   $\overline{\overline{X}} = \dfrac{5(13) + 6(9.83) + 8(4.375)}{19} = \dfrac{158.98}{19} = 8.367$, and

$\widehat{\sigma}^2_{\text{between}} = \dfrac{5(13-8.367)^2 + 6(9.83-8.367)^2 + 8(4.375-8.367)^2}{3-1} = \dfrac{247.654}{2} = 123.827$. Thus,
$\boldsymbol{F} = \dfrac{\widehat{\sigma}^2_{\text{between}}}{\widehat{\sigma}^2_{\text{within}}} = \dfrac{123.827}{24.169} = 5.12.$

Step 7: Since $\boldsymbol{F}$ falls in the rejection region, we reject H_0. At least one of the majors response mean is different.

4. Step 1: H_0: All population means are equal, and H_1: Not all population means are equal.
Step 2: $\alpha = 0.01$.
Step 3: We use an F distribution with $\text{df}_{\text{num}} = 4 - 1 = 3$, and $\text{df}_{\text{den}} = 28 - 4 = 24$.
Step 4: The critical F value is $F_{3,\,24,\,\alpha=0.01} = 4.72$.
Step 5: Reject H_0 in favor of H_1 if $\boldsymbol{F} > 4.72$. Otherwise, fail to reject H_0.
Step 6: The sample means are 14.429, 15.429, 19.143, and 15.143. The sum of the d^2 values is 343.1, so $\widehat{\sigma}^2_{\text{within}} = \dfrac{343.1}{24} = 14.3$. For $\widehat{\sigma}^2_{\text{between}}$, $\overline{\overline{X}} = 16.036$, and $\widehat{\sigma}^2_{\text{between}} = \dfrac{93.8}{3} = 31.3$. Thus, $\boldsymbol{F} = \dfrac{31.3}{14.3} = 2.19$.
Step 7: Since $\boldsymbol{F}$ does not fall in the rejection region, we fail to reject H_0. There is no significant difference in accuracy.

6. Step 1: H_0: All population means are equal, and H_1: Not all population means are equal.
Step 2: $\alpha = 0.05$.
Step 3: We'll use an F distribution with $\text{df}_{\text{num}} = 3 - 1 = 2$, and $\text{df}_{\text{den}} = 12 - 3 = 9$.
Step 4: The critical F value is $F_{2,\,9,\,\alpha=0.05} = 4.26$.
Step 5: Reject H_0 in favor of H_1 if $\boldsymbol{F} > 4.26$. Otherwise, fail to reject H_0.
Step 6: The sum of the d^2 values is $108,336$, so $\widehat{\sigma}^2_{\text{within}} = \dfrac{108,336}{9} = 12,037$. And $\widehat{\sigma}^2_{\text{between}} = \dfrac{64,586}{2} = 32,393$. Thus, $\boldsymbol{F} = \dfrac{32,393}{12,037} = 2.68$.
Step 7: Since $\boldsymbol{F}$ does not fall in the rejection region, we fail to reject H_0 and Ms. Breen's claim that the viewing habits are the same.

8. Step 1: H_0: All population means are equal, and H_1: Not all population means are equal.
Step 2: $\alpha = 0.01$.
Step 3: We'll use an F distribution with $\text{df}_{\text{num}} = 6 - 1 = 5$, and $\text{df}_{\text{den}} = 24 - 6 = 18$.
Step 4: The critical F value is $F_{5,\,18,\,\alpha=0.01} = 4.25$.
Step 5: Reject H_0 in favor of H_1 if $\boldsymbol{F} > 4.25$. Otherwise, fail to reject H_0.
Step 6: The sum of the d^2 values is $237,996$, so $\widehat{\sigma}^2_{\text{within}} = \dfrac{237,996}{18} = 13,222$. And $\widehat{\sigma}^2_{\text{between}} = \dfrac{313,185}{5} = 62,637$. Thus, $\boldsymbol{F} = \dfrac{62,637}{13,222} = 4.74$.
Step 7: Since $\boldsymbol{F}$ falls in the rejection region, we reject H_0. At least one of the computer sorting algorithms means is different.

10. Step 1: H_0: All population means are equal, and H_1: Not all population means are equal.
Step 2: $\alpha = 0.05$.
Step 3: We'll use an F distribution with $\text{df}_{\text{num}} = 3 - 1 = 2$, and $\text{df}_{\text{den}} = 12 - 3 = 9$.
Step 4: The critical F value is $F_{2,\,9,\,\alpha=0.05} = 4.26$.
Step 5: Reject H_0 in favor of H_1 if $\boldsymbol{F} > 4.26$. Otherwise, fail to reject H_0.

Step 6: The sum of the d^2 values is 35,462.8, so $\hat{\sigma}^2_{\text{within}} = \dfrac{35,462.8}{9} = 3,940.3$. And $\hat{\sigma}^2_{\text{between}} = \dfrac{42,930}{2} = 21,465$. Thus, $\boldsymbol{F} = \dfrac{21,465}{3,940.3} = 5.45$.

Step 7: Since $\boldsymbol{F}$ falls in the rejection region, we reject H_0, there is a significant difference in the mean number of unpopped kernels.

10-3 The One-Way ANOVA Table and Computers to the Rescue

2. "Error" is the natural variation of the members *within* the same population.

4. The SS error is the numerator for the $\hat{\sigma}^2_{\text{within}}$ estimate of σ^2. It is found by computing the mean for each of the four groups or factors, finding the deviation between each value and its corresponding mean, squaring the deviations, and then adding these squared deviations (see the procedure spelled out in the numerator of formula 10.4).

6. $\hat{\sigma}^2_{\text{within}}$

8. The test statistic $\boldsymbol{F} = \dfrac{\text{MS factor}}{\text{MS error}}$. In this case, $\dfrac{407}{467} = 0.87$.

10. "Level" refers to the different factors or treatments.

12. We see how many factors there are, the number of values in each, the mean and standard deviation for each sample, and a confidence interval display.

14. With four factors, the $\text{df}_{\text{num}} = 4 - 1 = 3$. With a total of 16 values within the factor groups, $\text{df}_{\text{den}} = 16 - 4 = 12$.

16. With a p-value of 0.002, we reject H_0. There is only a 0.002 area under the F curve to the right of $F = 8.92$. $(0.002 < 0.05)$

18. There are 4 factors. Refer to the chart where the levels are listed.

20. The $\text{df}_{\text{num}} = 4 - 1 = 3$. And $\text{df}_{\text{den}} = 17 - 4 = 13$.

22. To compute the test statistic, $\boldsymbol{F}$, divide MS factor by MS error. So $\boldsymbol{F} = \dfrac{163.9}{25.0} = 6.56$.

24. Because we reject H_0, we would use Fisher's procedure to see which pairs of means are significantly different.

26. The MS error is the SS error divided by the DF error. In this case, MS error $= \dfrac{738.0}{8} = 92.3$. The MS factor is the SS factor divided by the DF factor. In this case, MS factor $= \dfrac{2092.9}{3} = 697.6$.

28. Since the p-value is 0.010, and since $0.010 < 0.05$, we reject H_0 at the 0.05 level.

30. The DF factor $= 3 - 1 = 2$. And the DF error is $14 - 3 = 11$.

32. To compute the test statistic, $\boldsymbol{F}$, divide MS factor by MS error. So $\boldsymbol{F} = \dfrac{96}{258} = 0.37$.

34. Because we failed to reject H_0, there is no need to compare pairs of means.

36. $B = 73.6 + 816.1 = 889.7$.

38. $D = \dfrac{816.1}{12} = 68$.

40. Fail to reject H_0 at the 0.01 level.

42. $B = 1{,}090.9$.

44. $D = 60.6$.

46. Reject H_0 at the 0.01 level.

48. Step 1: H_0: All population means are equal, and H_1: Not all population means are equal.
Step 2: $\alpha = 0.01$.
Step 3: We'll use an F distribution with $df_{num} = 5 - 1 = 4$, and $df_{den} = 15 - 5 = 10$.
Step 4: The critical F value is $F_{4,10,0.01} = 5.99$.
Step 5: Reject H_0 in favor of H_1 if the p-value is less than 0.01. Otherwise, fail to reject H_0.
Step 6: From the ANOVA output, the p-value is 0.000
Step 7: Since the p-value < 0.01, we reject H_0, there is a significant difference in the mean compressive strength of the different concrete mixes. (Also note from the ANOVA output that $\boldsymbol{F} = 58.89$.)

50.
```
One-way Analysis of Variance for Polymer

Analysis of Variance
Source     DF        SS        MS        F       P
Factor      3     48901     16300     4.70   0.007
Error      36    124796      3467
Total      39    173697
```

52. $\mu_{0.0\text{ cc}} \neq \mu_{0.5\text{ cc}}$, $\mu_{0.0\text{ cc}} \neq \mu_{1.0\text{ cc}}$, and $\mu_{0.0\text{ cc}} \neq \mu_{2.0\text{ cc}}$.

54. Step 1: H_0: All population means are equal, and H_1: Not all population means are equal.
Step 2: $\alpha = 0.01$.
Step 3: We'll use an F distribution with $df_{num} = 3 - 1 = 2$, and $df_{den} = 15 - 3 = 12$.
Step 4: The critical F value is $F_{2,\,12,\,\alpha=0.01} = 6.93$.
Step 5: Reject H_0 in favor of H_1 if the p-value is less than 0.01. Otherwise, fail to reject H_0.
Step 6: From the ANOVA table, the p-value is 0.002
Step 7: Since the p-value < 0.01, we reject H_0, there is a significant difference in the mean reflectance of different sputtering machines. (Also note from the ANOVA output that $\boldsymbol{F} = 11.14$.)

56.
```
One-way Analysis of Variance

Analysis of Variance for C4
Source     DF        SS        MS        F       P
C5          2      7891      3946     9.75   0.000
Error      57     23056       404
Total      59     30948
```

58. $\mu_{\text{Check}} \neq \mu_{\text{Cash}}$ and $\mu_{\text{Credit}} \neq \mu_{\text{Cash}}$.

Answers to Even-Numbered Exercises

2. As df_{den} increases the critical F value decreases.

4. As df_{num} increases the critical F value decreases.

6. Step 1: H_0: All population means are equal, and H_1: Not all population means are equal.

Step 2: $\alpha = 0.01$.

Step 3: We will use a F distribution.

Step 4: $F_{3,13,0.01} = 5.74$.

Step 5: Reject H_0 in favor of H_1 if $\boldsymbol{F} > 5.74$, otherwise fail to reject H_0.

Step 6: Calculate the values to obtain $\boldsymbol{F}$.

Thirties			Fifties		
x_1	$x_1 - \overline{x}_1$	$(x_1 - \overline{x}_1)^2$	x_2	$x_2 - \overline{x}_2$	$(x_2 - \overline{x}_2)^2$
4	−1.8	3.24	3	−2	4
5	−0.8	0.64	5	0	0
6	0.2	0.04	3	−2	4
8	2.2	4.84	7	2	4
6	0.2	0.04	7	2	4
29		8.80	25		16

$\overline{x}_1 = \frac{29}{5} = 5.8,\ d_1^2 = 8.80$ $\qquad$ $\overline{x}_2 = \frac{25}{5} = 5.0,\ d_2^2 = 16$

Sixties			Seventies		
x_3	$x_3 - \overline{x}_3$	$(x_3 - \overline{x}_3)^2$	x_4	$x_4 - \overline{x}_4$	$(x_4 - \overline{x}_4)^2$
4	−1.5	2.25	5	0	0
7	1.5	2.25	3	−2	4
7	1.5	2.25	7	2	4
4	−1.5	2.25	15		8
22		9.00			

$\overline{x}_3 = \frac{22}{4} = 5.5,\ d_3^2 = 9.00$ $\qquad$ $\overline{x}_4 = \frac{15}{3} = 5.0,\ d_4^2 = 8$

$$\overline{\overline{X}} = \frac{5(5.8) + 5(5.0) + 4(5.5) + 3(5.0)}{5 + 5 + 4 + 3} = \frac{91}{17} = 5.3529.$$

$$\widehat{\sigma}^2_{\text{between}} = \frac{5(5.8 - 5.35290^2 + 5(5.0 - 5.3529)^2 + 4(5.5 - 5.3529)^2 + 3(5.0 - 5.3529)^2}{4 - 1}$$
$$= \frac{5(0.1999) + 5(0.1245) + 4(0.0216) + 3(0.1245)}{3} = \frac{2.0819}{3} = 0.6940.$$

$$\widehat{\sigma}^2_{\text{within}} = \frac{8.8 + 16 + 9 + 8}{17 - 4} = \frac{41.8}{13} = 3.2154.$$

$$\boldsymbol{F} = \frac{\widehat{\sigma}^2_{\text{between}}}{\widehat{\sigma}^2_{\text{within}}} = \frac{0.6940}{3.2154} = 0.2158.$$

Step 7: Since $\boldsymbol{F}$ is less than 5.74, we fail to reject H_0, the means for all 4 age groups are equal.

Placing the data into the MINITAB program we obtain the following output.

```
One-way Analysis of Variance

Analysis of Variance
Source     DF        SS        MS        F        P
Factor      3      2.08      0.69     0.22    0.884
Error      13     41.80      3.22
Total      16     43.88
```

```
                                       Individual 95% CIs For Mean
                                       Based on Pooled StDev
Level        N      Mean     StDev    --+---------+---------+---------+----
Thirties     5     5.800     1.483               (-----------*----------)
Fifties      5     5.000     2.000         (----------*-----------)
Sixties      4     5.500     1.732           (------------*------------)
Seventie     3     5.000     2.000    (--------------*--------------)
                                      --+---------+---------+---------+----
Pooled StDev =     1.793              3.0       4.5       6.0       7.5
```

8. Step 1: H_0: All population means are equal, and H_1: Not all population means are equal.
Step 2: $\alpha = 0.01$.
Step 3: We use a F distribution.
Step 4: $F_{2,12,0.01} = 6.93$.
Step 5: Reject H_0 and accept H_1 if $\boldsymbol{F} > 6.93$, otherwise fail to reject H_0.
Step 6: Calculate the values to obtain $\boldsymbol{F}$.

Eye Level			Waist Level			Knee Level		
x_1	$x_1 - \overline{x}_1$	$(x_1 - \overline{x}_1)^2$	x_2	$x_2 - \overline{x}_2$	$(x_2 - \overline{x}_2)^2$	x_3	$x_3 - \overline{x}_3$	$(x_3 - \overline{x}_3)^2$
98	−3.6	12.96	106	6.4	40.96	103	8.8	77.44
106	4.4	19.36	105	5.4	29.16	95	0.8	0.64
111	9.4	88.36	98	−1.6	2.56	87	−7.2	51.84
85	−16.6	275.56	93	−6.6	43.56	94	−0.2	0.04
108	6.4	40.96	96	−3.6	12.96	92	−2.2	4.84
508		437.20	498		129.20	471		134.80

$\overline{x}_1 = \dfrac{508}{5} = 101.6$ $\qquad \overline{x}_2 = \dfrac{498}{5} = 99.6$ $\qquad \overline{x}_3 = \dfrac{471}{5} = 94.2$

$d_1^2 = 437.20$ $\qquad d_2^2 = 129.20$ $\qquad d_3^2 = 134.80$

Thus $\overline{\overline{X}} = \dfrac{5(101.6) + 5(99.6) + 5(94.2)}{5+5+5} = \dfrac{1477}{15} = 98.47.$

$$\hat{\sigma}^2_{\text{between}} = \frac{5(101.6 - 98.47)^2 + 5(99.6 - 98.47)^2 + 5(94.2 - 98.47)^2}{3-1}$$
$$= \frac{5(9.797) + 5(1.277) + 5(18.233)}{2} = \frac{146.535}{2} = 73.2675.$$
$$\hat{\sigma}^2_{\text{within}} = \frac{437.20 + 129.20 + 134.80}{15-3} = \frac{701.2}{12} = 58.433.$$
$$\boldsymbol{F} = \frac{\hat{\sigma}^2_{\text{between}}}{\hat{\sigma}^2_{\text{within}}} = \frac{73.2675}{58.433} = 1.254$$

Step 7: Since $\boldsymbol{F}$ is less than 6.93, we fail to reject H_0, there is no significant difference in the population mean sales based on the shelf location.

Placing the data into the MINITAB program we obtain the following output.

```
One-way Analysis of Variance

Analysis of Variance
Source      DF        SS        MS        F        P
Factor       2     146.5      73.3     1.25    0.320
Error       12     701.2      58.4
Total       14     847.7
                                       Individual 95% CIs For Mean
                                       Based on Pooled StDev
Level        N      Mean     StDev    -------+---------+---------+---------
Eye Leve     5    101.60     10.45                (----------*----------)
Waist Le     5     99.60      5.68             (---------*----------)
Knee Lev     5     94.20      5.81     (----------*---------)
                                      -------+---------+---------+---------
Pooled StDev =      7.64                   91.0      98.0     105.0
```

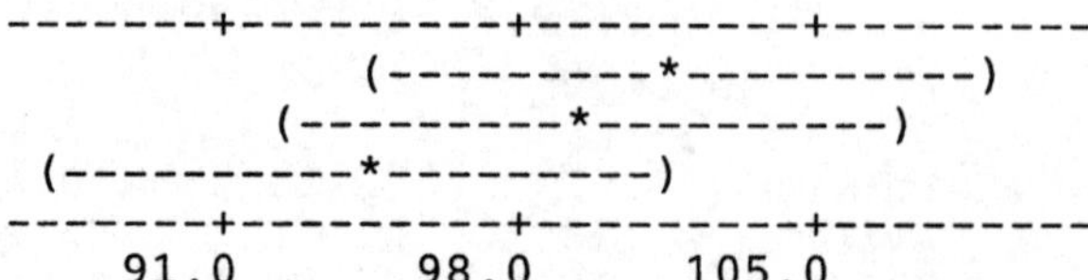

10. Step 1: H_0: All population means are equal, and H_1: Not all the population means are equal.
Step 2: $\alpha = 0.05$.
Step 3: We use a F distribution.
Step 4: $F_{2,12,0.05} = 3.89$.
Step 5: Reject H_0 and accept H_1 if $\boldsymbol{F} > 3.89$, otherwise fail to reject H_0.
Step 6: Calculate the values to obtain $\boldsymbol{F}$.

Brand A			Brand B			Brand C		
x_1	$x_1 - \overline{x}_1$	$(x_1 - \overline{x}_1)^2$	x_2	$x_2 - \overline{x}_2$	$(x_2 - \overline{x}_2)^2$	x_3	$x_3 - \overline{x}_3$	$(x_3 - \overline{x}_3)^2$
32.5	4.9	24.01	27.5	6.52	42.510	15.5	−0.06	0.0036
27.7	0.1	0.01	27.7	6.72	45.158	12.5	−3.06	9.3636
18.4	−9.2	84.64	30.3	9.32	86.862	18.9	0.34	11.1556
37.7	10.1	102.01	13.7	−7.28	52.998	15.8	0.24	0.0576
21.7	−5.9	34.81	5.7	−15.28	233.478	15.1	−0.46	0.2116
138.0		245.48	104.9		461.006	77.8		20.7920

$\overline{x}_1 = \dfrac{138}{5} = 27.6$ $\qquad \overline{x}_2 = \dfrac{104.9}{5} = 21.0$ $\qquad \overline{x}_3 = \dfrac{77.8}{5} = 15.56$

$d_1^2 = 245.48$ $\qquad d_2^2 = 461.006$ $\qquad d_3^2 = 20.792$

$$\overline{\overline{X}} = \frac{5(27.6) + 5(21.0) + 5(15.56)}{5+5+5} = \frac{320.7}{15} = 21.38.$$

$$\widehat{\sigma}^2_{\text{between}} = \frac{5(27.6 - 21.38)^2 + 5(21.0 - 21.38)^2 + 5(15.56 - 21.38)^2}{3-1}$$
$$= \frac{5(38.6884) + 5(0.1444) + 5(33.8724)}{2} = \frac{363.526}{2} = 181.763.$$

$$\widehat{\sigma}^2_{\text{within}} = \frac{245.48 + 461.006 + 20.792}{15-3} = \frac{727.278}{12} = 60.6065.$$

$$\boldsymbol{F} = \frac{\widehat{\sigma}^2_{\text{between}}}{\widehat{\sigma}^2_{\text{within}}} = \frac{181.763}{60.6065} = 2.999.$$

Step 7: Since $\boldsymbol{F}$ is less than 3.89, we fail to reject H_0, there is no significant difference in the population mean sales of these three brands.

Placing the data into the MINITAB program we obtain the following output.

```
One-way Analysis of Variance

Analysis of Variance
Source       DF        SS          MS         F        P
Factor        2     363.6       181.8      3.00    0.088
Error        12     727.3        60.6
Total        14    1090.9
                                       Individual 95% CIs For Mean
                                       Based on Pooled StDev
Level         N      Mean       StDev  -+---------+---------+---------+-----
Brand A       5    27.600       7.834                    (---------*--------)
Brand B       5    20.980      10.736              (--------*---------)
Brand C       5    15.560       2.280    (--------*---------)
                                       -+---------+---------+---------+-----
Pooled StDev =      7.785             8.0      16.0      24.0      32.0
```

12. Step 1: H_0: All population means are equal, and H_1: Not all the population means are equal.
Step 2: $\alpha = 0.01$.
Step 3: We use a F distribution.
Step 4: $F_{2,17,0.01} = 6.11$.
Step 5: Reject H_0 and accept H_1 if $\boldsymbol{F} > 6.11$, otherwise fail to reject H_0.
Step 6: Calculate the values to obtain $\boldsymbol{F}$.

Nicodependent			Nondependent			Nonsmokers		
x_1	$x_1 - \overline{x}_1$	$(x_1 - \overline{x}_1)^2$	x_2	$x_2 - \overline{x}_2$	$(x_2 - \overline{x}_2)^2$	x_3	$x_3 - \overline{x}_3$	$(x_3 - \overline{x}_3)^2$
87	1.5	2.25	58	−5.86	34.34	86	12.71	161.544
141	55.5	3080.25	50	−13.86	192.10	68	−5.29	27.984
128	42.5	1806.25	44	−19.86	394.42	72	−1.29	1.664
63	−22.5	506.25	97	33.14	1098.26	63	−10.29	105.884
47	−38.5	1482.25	80	16.14	260.50	79	5.71	32.604
47	−38.5	1482.25	63	−0.86	0.74	73	−0.29	0.084
513		8359.50	55	−8.86	78.50	72	−1.29	1.664
			447		2058.86	513		331.428

$\overline{x}_1 = \frac{513}{6} = 85.5$ $\overline{x}_2 = \frac{447}{7} = 63.86$ $\overline{x}_3 = \frac{513}{7} = 73.29$

$d_1^2 = 8359.50$ $d_2^2 = 2058.86$ $d_3^2 = 331.428$

$$\overline{\overline{X}} = \frac{6(85.5) + 7(63.86) + 7(73.29)}{6+7+7} = \frac{1473}{20} = 73.65$$

$$\widehat{\sigma}^2_{\text{between}} = \frac{6(85.5 - 73.65)^2 + 7(63.86 - 73.65)^2 + 7(73.29 - 73.65)^2}{3-1}$$

$$= \frac{6(140.423)+7(95.844)+7(0.130)}{2} = \frac{1514.356}{2} = 757.178.$$

$$\widehat{\sigma}^2_{\text{within}} = \frac{8359.50 + 2058.86 + 331.428}{20-3} = \frac{10749.788}{17} = 632.340.$$

$$F = \frac{\widehat{\sigma}^2_{\text{between}}}{\widehat{\sigma}^2_{\text{within}}} = \frac{757.178}{632.340} = 1.197.$$

Step 7: Since F is less than 6.11, we fail to reject H_0, there is no significant difference in the population mean emotional distress scores

Placing the data into the MINITAB program we obtain the following output.

```
One-way Analysis of Variance

Analysis of Variance
Source     DF        SS         MS        F        P
Factor      2      1515        757     1.20    0.326
Error      17     10750        632
Total      19     12265
                                   Individual 95% CIs For Mean
                                   Based on Pooled StDev
Level       N      Mean      StDev  ---------+---------+---------+-------
Nico-Dep    6     85.50      40.89            (----------*----------)
Nondepen    7     63.86      18.52   (---------*---------)
Nonsmoke    7     73.29       7.43        (---------*---------)
                                    ---------+---------+---------+-------
Pooled StDev =    25.15                     60        80       100
```

14. Factor is the difference between the populations. The factor variation is also called the treatment.

16. The df of the factor is the number of sample items minus one and is the denominator for $\widehat{\sigma}^2_{\text{between}}$. In this example ANOVA output this is $4-1$. The df of the error is the total number of sample items minus the number of samples and is the denominator for $\widehat{\sigma}^2_{\text{within}}$. In this example ANOVA output this is, $(5+5+4+3) - 4 = 13$. The total df is the total number of sample items minus one.

18. The SS factor is the numerator for the $\widehat{\sigma}^2_{\text{between}}$ estimate of σ^2. It is found by computing the mean for each of the four groups or factors and then computing the grand mean, finding the deviation between each of the factor means and the grand mean, squaring the deviations, multiplying each squared deviation by the number of samples in the factor and then adding these values (see the procedure spelled out in the numerator of formula 10.6).

20. The MS factor is the $\frac{\text{SS factor}}{\text{DF factor}}$, which is $\hat{\sigma}^2_{\text{between}}$.

22. The p-value is the area to the right of the test statistic. In this example, $p = 0.884$ and since $0.884 > 0.05$ we fail to reject H_0.

24. The bottom half of the output gives each level's mean, sample size, standard deviation, and the 95% confidence interval for each using the pooled standard deviation.

26. The pooled standard deviation $= \sqrt{\hat{\sigma}^2_{\text{within}}}$.

28. $F = \frac{\text{MS factor}}{\text{MS factor}}$. In this case, $F = \frac{31.18}{5.27} = 5.92$.

30. Because we rejected H_0, we would use Fisher's procedure to see which pairs of means are significantly different

32. Using the SS column, B $= 1515 + 10750 = 12265$.

34. In Exercise 31 we found A $= 17$. Now using the ERROR row, D $= \frac{\text{SS error}}{\text{DF error}} = \frac{10750}{17} = 632.4$.

36. Since $F_{2,17,0.05} = 3.59 > F = 1.20$, we fail to reject H_0.

38. Using the SS column, B $+ 43.33 = 51.83$, so B $= 51.83 - 43.33 = 8.50$.

40. D $= \frac{43.33}{20} = 2.17$.

42. Since $F_{3,20,0.01} = 4.94 > F = 1.31$, we fail to reject H_0.

44. Step 1: H_0: All population means are equal, and H_1: Not all population means are equal.
Step 2: $\alpha = 0.01$.
Step 3: We'll use an F distribution with $df_{num} = 2$, and $df_{den} = 24$.
Step 4: The critical F value is $F_{2,\,24,\,\alpha=0.01} = 5.61$.
Step 5: Reject H_0 in favor of H_1 if the p-value is less than 0.01. Otherwise, fail to reject H_0.
Step 6: From the ANOVA table, the p-value is 0.261
Step 7: Since the p-value > 0.01, we fail to reject H_0, there is a not significant difference in the mean widths of buckwheat.

46.
```
One-way Analysis of Variance

Analysis of Variance for Factor
Source      DF        SS        MS        F        P
Factor       3     60.80     20.27     6.27    0.001
Error       60    193.94      3.23
Total       63    254.73
```

48. $\mu_{\text{Marathon}} \neq \mu_{\text{1 Tab}}$, $\mu_{\text{Marathon}} \neq \mu_{\text{2 Tab}}$, and $\mu_{\text{Marathon}} \neq \mu_{\text{Control}}$

50. Step 1: H_0: All population means are equal, and H_1: Not all population means are equal.
Step 2: $\alpha = 0.01$.
Step 3: We'll use an F distribution with $df_{num} = 5$, and $df_{den} = 12$.

Step 4: The critical F value is $F_{5,\,12,\,\alpha=0.01} = 5.06$.

Step 5: Reject H_0 in favor of H_1 if the p-value is less than 0.01. Otherwise, fail to reject H_0.

Step 6: From the ANOVA table, the p-value is 0.000

Step 7: Since the p-value < 0.01, we reject H_0, there is a significant difference in the mean strengths of the various concentrations of fly ash.

Chapter 11 Chi-Square Test: Goodness-of-Fit and Contingency Table Methods

Answers to Even-Numbered Self-Testing Review

11-1 Chi-Square Testing: Purpose and Procedure

2. The objective is to test if an observed distribution is representative of one that is spelled out in the null hypothesis.

4. The objective is to see if two variables are independent. That is, to see if the occurrence (or non-occurrence) of one does not affect the probability of occurrence of the other.

6. The level of significance and the degrees of freedom.

8. There are $4 - 1 = 3$ degrees of freedom, and the critical χ^2 value at the 0.01 level is 11.34.

10. The degrees of freedom $= (2 - 1)(3 - 1) = (1)(2) = 2$, and the critical χ^2 value at the 0.05 level is 5.99.

11-2 The Goodness-of-Fit Test

2. Step 1: H_0: There has been no change in the distribution of age groups. That is, there are still 10, 13, 32, and 45 percent of the residents in the 4 respective age categories. H_1: The age distribution has changed since 1987.
   Step 2: $\alpha = 0.05$.
   Step 3: χ^2.
   Step 4: The critical χ^2 value with $4 - 1 = 3$ df is 7.81.
   Step 5: Reject H_0 in favor of H_1 if $\chi^2 > 7.81$. Otherwise, fail to reject H_0.
   Step 6: $\chi^2 = \sum\left[\frac{(O-E)^2}{E}\right] = 105.856 + 45.847 + 8.304 + 36.717 = 196.724$.
   Step 7: Since χ^2 falls in the rejection region, we reject H_0. There has been a significant change in the age distribution.

4. Step 1: H_0: The distribution follows a 9:3:3:1 ratio. H_1: The distribution does not follow such a ratio.
   Step 2: $\alpha = 0.05$.
   Step 3: χ^2.

Step 4: The critical χ^2 value with 4 – 1 or 3 df is 7.81.
Step 5: Reject H_0 in favor of H_1 if $\chi^2 > 7.81$. Otherwise, fail to reject H_0.
Step 6: $\chi^2 = \sum\left[\frac{(O-E)^2}{E}\right] = 1.788 + 8.333 + 0.000 + 1.000 = 11.11$.
Step 7: Since χ^2 falls in the rejection region, we reject H_0. We conclude at the 0.05 level that the distribution does not follow the Mendelian ratio.

6. Step 1: H_0: The distribution is uniform--we should expect $\frac{1}{7}$ of the crimes to be committed each day. H_1: The distribution isn't uniform.
Step 2: $\alpha = 0.05$.
Step 3: χ^2.
Step 4: The critical χ^2 value with $7 - 1 = 6$ df is 12.59.
Step 5: Reject H_0 in favor of H_1 if $\chi^2 > 12.59$. Otherwise, fail to reject H_0.
Step 6: $\chi^2 = \sum\left[\frac{(O-E)^2}{E}\right] = 35.660$.
Step 7: Since a χ^2 falls in the rejection region, we reject H_0. The crime distribution isn't uniform.

8. Step 1: H_0: The distribution fits the dean of student's distribution. That is, the expected percentages in the three categories are 50, 33, and 17. H_1: The distribution doesn't fit the one spelled out by the dean.
Step 2: $\alpha = 0.05$.
Step 3: χ^2.
Step 4: The critical χ^2 value with $3 - 1 = 2$ df is 5.99.
Step 5: Reject H_0 in favor of H_1 if$\chi^2 > 5.99$. Otherwise, fail to reject H_0.
Step 6: $\chi^2 = \sum\left[\frac{(O-E)^2}{E}\right] = 3.33 + 1.25 + 2.50 = 7.08$.
Step 7: Since a χ^2 falls in the rejection region, we reject H_0. The sample distribution doesn't fit the one spelled out in the dean.

10. Step 1: H_0: There has been no change in the distribution of sexes. That is, there are still 56 percent males and 44 percent females. H_1: The age distribution has changed since 1991.
Step 2: $\alpha = 0.05$.
Step 3: χ^2.
Step 4: The critical χ^2 value with $2 - 1 = 1$ df is 3.84.
Step 5: Reject H_0 in favor of H_1 if $\chi^2 > 3.84$. Otherwise, fail to reject H_0.
Step 6: $\chi^2 = \sum\left[\frac{(O-E)^2}{E}\right] = 0.0095 + 0.0012 = 0.0107$.
Step 7: Since χ^2 does not fall in the rejection region, we fail to reject H_0. There has not been a significant change in the sex distribution.

12. Step 1: H_0: The distribution is uniform--we should expect $\frac{1}{3}$ of the students to work out in each time periods. H_1: The distribution isn't uniform.
Step 2: $\alpha = 0.05$.
Step 3: χ^2.
Step 4: The critical χ^2 value with $3 - 1 = 2$ df is 5.99.
Step 5: Reject H_0 in favor of H_1 if $\chi^2 > 5.99$. Otherwise, fail to reject H_0.
Step 6: $\chi^2 = \sum\left[\frac{(O-E)^2}{E}\right] = 5.400 + 0.267 + 8.067 = 13.734$.

Step 7: Since a χ^2 falls in the rejection region, we reject H_0. The workout times are not preferred uniform.

14. Step 1: H_0: The physical activity frequency distribution of the college females is the same as working women. That is, the expected percentages in the six categories are 12.5, 37.2, 22.1, 11.5, 10.8, and 5.9. H_1: The physical activity frequency distribution is different.
Step 2: $\alpha = 0.05$.
Step 3: χ^2.
Step 4: The critical χ^2 value with $6 - 1 = 5$ df is 11.07.
Step 5: Reject H_0 in favor of H_1 if $\chi^2 > 11.07$. Otherwise, fail to reject H_0.
Step 6: $\chi^2 = \sum\left[\frac{(O-E)^2}{E}\right] = 3.38 + 2.08 + 16.51 + 3.67 + 11.61 + 0.14 = 37.39.$
Step 7: Since χ^2 falls in the rejection region, we reject H_0. The physical activity of college women is different from the surveyed 1000 working women.

11-3 The Contingency Table Test

2. Step 1: H_0: Religion and political party of students are independent. H_1: Religion and political party of students arc not independent.
Step 2: $\alpha = 0.05$.
Step 3: χ^2.
Step 4: The critical χ^2 value with 6 df is 12.59.
Step 5: Reject H_0 in favor of H_1 if $\chi^2 > 12.59$. Otherwise, fail to reject H_0.
Step 6: From the MINITAB output, $\chi^2 = 50.330$.
Step 7: Since χ^2 falls in the rejection region, we reject H_0. Religion and political party of students are not independent.

4. Step 1: H_0: There's an equal percentage of male and female inner-city middle school students who want to stay in school. H_1: The variables are not independent.
Step 2: $\alpha = 0.05$.
Step 3: χ^2.
Step 4: The critical χ^2 value with $(2-1)(2-1) = 1$ df is 3.84.
Step 5: Reject H_0 in favor of H_1 if $\chi^2 > 3.84$. Otherwise, fail to reject H_0.
Step 6: We produce the following contingency table.

Row-Column	O	E	$O-E$	$(O-E)^2$	$\frac{(O-E)^2}{E}$
$1-1$	76	56.5	19.5	380.25	6.730
$1-2$	118	137.5	-19.5	380.25	2.765
$2-1$	37	56.5	-19.5	380.25	6.730
$2-2$	157	137.5	19.5	380.25	2.765
	388	388.0	0		18.991

So $\chi^2 = 18.991$.
Step 7: Since χ^2 falls in the rejection region, we reject H_0. Gender and inner-city middle school students who want to stay in school are not independent.

6. Step 1: H_0: The opinion about Clinton's budget is independent of income level. H_1: The opinion is not independent of income.
Step 2: $\alpha = 0.05$.
Step 3: χ^2.

Step 4: The critical χ^2 value with $(4-1)(4-1) = 9$ df is 16.92.
Step 5: Reject H_0 in favor of H_1 if $\chi^2 > 16.92$. Otherwise, fail to reject H_0.
Step 6: We produce the following (abbreviated) contingency table.

Row-Column	O	E	$\frac{(O-E)^2}{E}$	Row-Column	O	E	$\frac{(O-E)^2}{E}$
1 – 1	65	70.44	0.420	3 – 1	180	162.95	1.784
1 – 2	30	42.02	3.440	3 – 2	111	97.22	1.955
1 – 3	50	31.56	10.781	3 – 3	38	73.00	16.779
1 – 4	5	5.98	0.161	3 – 4	18	13.84	1.251
2 – 1	154	157.31	0.070	4 – 1	72	80.30	0.858
2 – 2	77	93.85	3.026	4 – 2	63	47.91	4.755
2 – 3	90	70.47	5.410	4 – 3	33	35.97	0.246
2 – 4	14	13.36	0.031	4 – 4	3	6.82	2.139
	485		23.339		518		29.767

So $\chi^2 = 23.339 + 29.767 = 53.106$.
Step 7: Since χ^2 falls in the rejection region, we reject H_0. Opinion about the budget and income level are not independent.

8. Step 1: H_0: The time from injury to surgery and cartilage condition are independent. H_1: The variables are not independent.
Step 2: $\alpha = 0.05$.
Step 3: χ^2.
Step 4: The critical χ^2 value with $(3-1)(3-1) = 4$ df is 9.49.
Step 5: Reject H_0 in favor of H_1 if $\chi^2 > 9.49$. Otherwise, fail to reject H_0.
Step 6: Hold everything! There's a problem here--we've run into a situation that violates one of the assumptions for contingency table tests that was spelled out early in the chapter. This assumption was that the expected frequency for each cell in the contingency table must be at least 5. But you'll note that the expected frequency for the cell at the intersection of the "Subacute" row and "Normal cartilage" column is only 3.337. Thus, this test may not produce valid results. When this is the case, the analyst can either combine some categories or try for a larger sample.

10. Step 1: H_0: The college in which a student is enrolled in is independent of whether or not the student had participated in high school sports. H_1: The variables are not independent.
Step 2: $\alpha = 0.05$.
Step 3: χ^2.
Step 4: The critical χ^2 value with $(5-1)(2-1) = 4$ df is 9.49.
Step 5: Reject H_0 in favor of H_1 if $\chi^2 > 9.49$. Otherwise, fail to reject H_0.
Step 6: We produce the following (abbreviated) contingency table.

Row-Column	O	E	$\frac{(O-E)^2}{E}$	Row-Column	O	E	$\frac{(O-E)^2}{E}$
1 – 1	58	38.69	9.638	3 – 2	28	32.90	0.730
1 – 2	15	34.31	10.868	4 – 1	32	37.63	0.842
2 – 1	51	54.59	0.236	4 – 2	39	33.37	0.950
2 – 2	52	48.41	0.266	5 – 1	29	43.99	5.108
3 – 1	42	37.10	0.647	5 – 2	54	39.01	5.760
	218		21.655		182		13.390

So $\chi^2 = 21.655 + 13.390 = 35.045$.
Step 7: Since χ^2 falls in the rejection region, we reject H_0. The college in which a student is enrolled in *and* whether or not the student had participated in high school sports are not independent.

12. Step 1: H_0: The answer to the question is independent of gender. H_1: The answer to the question is not independent of gender.
Step 2: $\alpha = 0.10$.
Step 3: χ^2.
Step 4: The critical χ^2 value with $(2-1)(2-1) = 1$ df is 2.71.
Step 5: Reject H_0 in favor of H_1 if $\chi^2 > 2.71$. Otherwise, fail to reject H_0.
Step 6: We produce the following contingency table.

Row-Column	O	E	$O-E$	$(O-E)^2$	$\frac{(O-E)^2}{E}$
1 − 1	9	6.12	2.88	8.2944	1.349
1 − 2	5	7.87	−2.87	8.2369	1.050
2 − 1	5	7.87	−2.87	8.2369	1.050
2 − 2	13	10.12	2.88	8.2944	0.816
	32	31.98	0.02		4.265

So $\chi^2 = 4.265$.
Step 7: Since χ^2 falls in the rejection region, we reject H_0. The answer to the question is not independent of gender.

14. Step 1: H_0: The location in the restaurant is independent of the type of alcohol consumed. H_1: The location in the restaurant and the type of alcohol consumed are not independent.
Step 2: $\alpha = 0.01$.
Step 3: χ^2.
Step 4: The critical χ^2 value with $(2-1)(3-1) = 2$ df is 9.21.
Step 5: Reject H_0 in favor of H_1 if $\chi^2 > 9.21$. Otherwise, fail to reject H_0.
Step 6: We produce the following contingency table.

Row-Column	O	E	$O-E$	$(O-E)^2$	$\frac{(O-E)^2}{E}$
1 − 1	18	19.33	−1.33	1.769	0.092
1 − 2	26	15.19	10.81	116.856	7.693
1 − 3	8	17.49	−9.49	90.060	5.149
2 − 1	24	22.67	1.33	1.769	0.078
2 − 2	7	17.81	−10.81	116.856	6.561
2 − 3	30	20.51	9.49	90.060	4.391
	113	113			23.964

So $\chi^2 = 23.964$.
Step 7: Since χ^2 falls in the rejection region, we reject H_0. The location in the restaurant and the type of alcohol consumed are not independent.

16. Step 1: H_0: The success or failure of the surgery is independent of prior surgical history. H_1: The variables are not independent.
Step 2: $\alpha = 0.05$.
Step 3: χ^2.
Step 4: The critical χ^2 value with $(2-1)(2-1) = 1$ df is 3.84.
Step 5: Reject H_0 in favor of H_1 if $\chi^2 > 3.84$. Otherwise, fail to reject H_0.
Step 6: We produce the following contingency table.

Row-Column	O	E	$O-E$	$(O-E)^2$	$\frac{(O-E)^2}{E}$
$1-1$	19	17.89	1.11	1.2321	0.069
$1-2$	31	32.14	-1.14	1.2996	0.040
$2-1$	11	12.14	-1.14	1.2996	0.107
$2-2$	23	21.86	1.14	1.2996	0.059
	84	84.03			0.275

So $\chi^2 = 0.275$.

Step 7: Since χ^2 does not fall in the rejection region, we fail to reject H_0. The rate of success was not significantly different for the two groups.

18. Step 1: H_0: The student use of marijuana is independent of the parental use of alcohol and drugs. H_1: The variables are not independent.

Step 2: $\alpha = 0.01$.

Step 3: χ^2.

Step 4: The critical χ^2 value with $(3-1)(3-1) = 4$ df is 13.28.

Step 5: Reject H_0 in favor of H_1 if $\chi^2 > 13.28$. Otherwise, fail to reject H_0.

Step 6: We produce the following contingency table.

Row-Column	O	E	$O-E$	$(O-E)^2$	$\frac{(O-E)^2}{E}$
$1-1$	141	119.35	21.65	468.723	3.927
$1-2$	68	82.78	-14.78	218.448	2.639
$1-3$	17	23.87	-6.87	47.197	1.977
$2-1$	54	57.56	-3.56	12.674	0.220
$2-2$	44	39.93	4.07	16.565	0.415
$2-3$	11	11.51	-0.51	0.260	0.023
$3-1$	40	58.09	-18.09	327.248	5.633
$3-2$	51	40.29	10.71	114.704	2.847
$3-3$	19	11.62	7.38	54.464	4.687
	445	445.0			22.368

So $\chi^2 = 22.368$.

Step 7: Since χ^2 falls in the rejection region, we reject H_0. The marijuana use is related to parental use of alcohol and drugs.

20. Step 1: H_0: For village children 1 to 6 years old, gender is independent of whether or not a person is affected by the disease. H_1: The variables are not independent.

Step 2: $\alpha = 0.05$.

Step 3: χ^2.

Step 4: The critical χ^2 value with $(2-1)(2-1) = 1$ df is 6.63.

Step 5: Reject H_0 in favor of H_1 if $\chi^2 > 6.63$. Otherwise, fail to reject H_0.

Step 6: We produce the following contingency table.

Row-Column	O	E	$O-E$	$(O-E)^2$	$\frac{(O-E)^2}{E}$
$1-1$	5	6.31	-1.31	1.7161	0.272
$1-2$	26	24.69	1.31	1.7161	0.070
$2-1$	7	5.69	1.31	1.7161	0.302
$2-2$	21	22.31	-1.31	1.7161	0.077
	39	39.0			0.721

So $\chi^2 = 0.721$.

Step 7: Since χ^2 does not fall in the rejection region, we fail to reject H_0. For village children 1 to 6 years old, gender is independent of whether or not a person is affected by the disease

22. Step 1: H_0: For village children 7 to 17 years old, gender is independent of whether or not a person is affected by the disease. H_1: The variables are not independent.
Step 2: $\alpha = 0.05$.
Step 3: χ^2.
Step 4: The critical χ^2 value with $(2-1)(2-1) = 1$ df is 6.63.
Step 5: Reject H_0 in favor of H_1 if $\chi^2 > 6.63$. Otherwise, fail to reject H_0.
Step 6: We produce the following contingency table.

Row-Column	O	E	$O-E$	$(O-E)^2$	$\frac{(O-E)^2}{E}$
$1-1$	328	325.11	2.89	8.3521	0.0257
$1-2$	132	134.89	−2.89	8.3521	0.0619
$2-1$	289	291.89	−2.89	8.3521	0.0286
$2-2$	124	121.11	2.89	8.3521	0.0690
	873	873.0			0.1870

So $\chi^2 = 0.187$.
Step 7: Since χ^2 does not fall in the rejection region, we fail to reject H_0. For village children 7 to 17 years old, gender is independent of whether or not a person is affected by the disease.

Answers to Even-Numbered Exercises

2. Step 1: H_0: The percentage of students able to do 40 or more pushups in 60 seconds at the end of a two-hour workout is independent of the gender. H_1: The ability to do 40 or more pushups in 60 seconds and gender are not independent.
Step 2: $\alpha = 0.05$.
Step 3: χ^2.
Step 4: The critical χ^2 value with $(2-1)(2-1) = 1$ df is 3.84.
Step 5: Reject H_0 in favor of H_1 if $\chi^2 > 3.84$. Otherwise, fail to reject H_0.
Step 6: We produce the following contingency table.

Row-Column	O	E	$O-E$	$(O-E)^2$	$\frac{(O-E)^2}{E}$
$1-1$	17	12.833	4.167	17.3639	1.3531
$1-2$	11	15.167	−4.167	17.3639	1.1449
$2-1$	5	9.167	−4.167	17.3639	1.8942
$2-2$	15	10.833	4.167	17.3639	1.6029
	48	48.0	0		5.9951

So $\chi^2 = 5.995$.
Step 7: Since χ^2 falls in the rejection region, we reject H_0. The ability to do 40 or more pushups in 60 seconds and gender are not independent.

4. Step 1: H_0: The felony type is independent of gender. H_1: The felony type is not independent of gender.
Step 2: $\alpha = 0.01$.
Step 3: χ^2.
Step 4: The critical χ^2 value with $(13-1)(2-1) = 12$ df is 26.2.
Step 5: Reject H_0 in favor of H_1 if $\chi^2 > 26.2$. Otherwise, fail to reject H_0.
Step 6: We produce the following contingency table.

Row-Column	O	E	$\frac{(O-E)^2}{E}$	Row-Column	O	E	$\frac{(O-E)^2}{E}$
1 − 1	18,507	18,145.4	7.2067	1 − 2	948	1,309.62	99.853
2 − 1	16,818	16,297.7	16.6085	2 − 2	656	1,176.27	230.118
3 − 1	14,562	14,198.3	9.318	3 − 2	661	1,024.74	129.113
4 − 1	2,042	1,909.2	9.2359	4 − 2	5	137.79	127.971
5 − 1	7,990	7,514.6	30.0703	5 − 2	67	542.36	416.637
6 − 1	2,007	1,940.9	2.2497	6 − 2	74	140.08	31.172
7 − 1	14,664	14,471.5	2.5596	7 − 2	852	1,044.46	35.464
8 − 1	11,543	12,212.6	36.7100	8 − 2	1,551	881.43	508.633
9 − 1	4,686	4,530.1	5.3687	9 − 2	171	326.95	74.386
10 − 1	1,382	1,709.6	62.7794	10 − 2	451	123.39	869.830
11 − 1	465	523.2	6.4825	11 − 2	96	37.76	89.828
12 − 1	37,131	38,611.3	56.7510	12 − 2	4,267	2,786.72	786.311
13 − 1	9,709	9,441.6	7.5749	13 − 2	414	681.43	104.954
	141,506	141,506.0	252.92		10,213	10,213.0	3,504.3

So $\chi^2 = 252.92 + 3,504.3 = 3,757.22$.

Step 7: Since χ^2 falls in the rejection region, we reject H_0. The felony type is not independent of gender

6. Step 1: H_0: There's no difference in the reactions of the workers in the three locals to the proposed change. H_1: The worker's opinion is not independent of the local they belong to.
Step 2: $\alpha = 0.05$.
Step 3: χ^2.
Step 4: The critical χ^2 value with $(3-1)(3-1) = 4$ df is 9.49.
Step 5: Reject H_0 in favor of H_1 if $\chi^2 > 9.49$. Otherwise, fail to reject H_0.
Step 6: We produce the following contingency table.

Row-Column	O	E	$O-E$	$(O-E)^2$	$\frac{(O-E)^2}{E}$
1 − 1	17	15	2	4	0.267
1 − 2	23	20	3	9	0.450
1 − 3	10	15	−5	25	1.667
2 − 1	9	9	0	0	0.000
2 − 2	13	12	1	1	0.083
2 − 3	8	9	−1	1	0.111
3 − 1	4	6	−2	4	0.667
3 − 2	4	8	−4	16	2.000
3 − 3	12	6	6	36	6.000
	100	100.0			11.244

So $\chi^2 = 11.244$.

Step 7: Since χ^2 falls in the rejection region, we reject H_0. The worker's opinion is not independent of the local they belong to.

8. Step 1: H_0: The proportion of concordance for the handedness trait is the same for identical and fraternal twins. H_1: The proportion of concordance for the handedness trait is different for the different type of twins.
Step 2: $\alpha = 0.01$.
Step 3: χ^2.
Step 4: The critical χ^2 value with $(2-1)(2-1) = 1$ df is 6.63.
Step 5: Reject H_0 in favor of H_1 if $\chi^2 > 6.63$. Otherwise, fail to reject H_0.
Step 6: We produce the following contingency table.

Row-Column	O	E	$O-E$	$(O-E)^2$	$\frac{(O-E)^2}{E}$
$1-1$	107	104.61	2.39	5.7121	0.054
$1-2$	28	20	−2.39	5.7121	0.187
$2-1$	406	9	−2.39	5.7121	0.014
$2-2$	121	12	2.39	5.7121	0.048
	662	662.0			0.304

So $\chi^2 = 0.304$.

Step 7: Since χ^2 does not fall in the rejection region, we fail to reject H_0. The proportion of concordance for the handedness trait is the same for identical and fraternal twins.

10. Step 1: H_0: The primary reading language is independent of the student's grade. H_1: The primary reading language and the grade of the student are not independent.

Step 2: $\alpha = 0.01$.

Step 3: χ^2.

Step 4: The critical χ^2 value with $(2-1)(2-1) = 1$ df is 6.63.

Step 5: Reject H_0 in favor of H_1 if $\chi^2 > 6.63$. Otherwise, fail to reject H_0.

Step 6: We produce the following contingency table.

Row-Column	O	E	$O-E$	$(O-E)^2$	$\frac{(O-E)^2}{E}$
$1-1$	242	241.02	0.98	0.9604	0.004
$1-2$	293	293.98	−0.98	0.9604	0.003
$2-1$	295	295.98	−0.98	0.9604	0.003
$2-2$	362	361.02	0.98	0.9604	0.003
	1,192	1,192.0			0.013

So $\chi^2 = 0.013$.

Step 7: Since χ^2 does not fall in the rejection region, we fail to reject H_0. The primary reading language is independent of the student's grade.

12. Step 1: H_0: The inpatient/outpatient status is independent of the criteria for admission. H_1: The inpatient/outpatient status and the criteria for admission are not independent.

Step 2: $\alpha = 0.05$.

Step 3: χ^2.

Step 4: The critical χ^2 value with $(4-1)(2-1) = 3$ df is 7.81.

Step 5: Reject H_0 in favor of H_1 if $\chi^2 > 7.81$. Otherwise, fail to reject H_0.

Step 6: We produce the following contingency table.

Row-Column	O	E	$O-E$	$(O-E)^2$	$\frac{(O-E)^2}{E}$
$1-1$	128	149.58	−21.58	465.696	3.113
$1-2$	88	66.42	21.58	465.696	7.011
$2-1$	103	89.33	13.67	186.869	2.092
$2-2$	26	39.67	−13.67	186.869	4.711
$3-1$	100	74.10	25.90	670.810	9.053
$3-2$	7	32.90	−25.90	670.810	20.389
$4-1$	196	213.99	−17.99	323.640	1.512
$4-2$	362	95.01	17.99	323.640	3.406
	761	761.0			51.287

So $\chi^2 = 51.287$.

Step 7: Since χ^2 falls in the rejection region, we reject H_0. The inpatient/outpatient status and the criteria for admission are not independent.

14. Step 1: H_0: The preferred vacation location is independent of gender. H_1: The preferred vacation location and gender are not independent.
Step 2: $\alpha = 0.05$.
Step 3: χ^2.
Step 4: The critical χ^2 value with $(2-1)(3-1) = 2$ df is 5.99.
Step 5: Reject H_0 in favor of H_1 if $\chi^2 > 5.99$. Otherwise, fail to reject H_0.
Step 6: We produce the following contingency table.

Row-Column	O	E	$O-E$	$(O-E)^2$	$\frac{(O-E)^2}{E}$
$1-1$	3	7.99	−4.99	24.9001	3.114
$1-2$	16	16.51	−0.51	0.2601	0.016
$1-3$	22	16.51	5.49	30.1401	1.828
$2-1$	12	7.01	4.99	24.9001	3.546
$2-2$	15	14.49	0.51	0.2601	0.018
$2-3$	9	14.49	−5.49	30.1401	2.082
	77	77.0			10.604

So $\chi^2 = 10.604$.
Step 7: Since χ^2 falls in the rejection region, we reject H_0. The preferred vacation location and gender are not independent.

16. Step 1: H_0: The age is independent of the side to which the saggital crest deflects. H_1: Age and the side to which the saggital crest deflects are not independent.
Step 2: $\alpha = 0.05$.
Step 3: χ^2.
Step 4: The critical χ^2 value with $(2-1)(3-1) = 2$ df is 5.99.
Step 5: Reject H_0 in favor of H_1 if $\chi^2 > 5.99$. Otherwise, fail to reject H_0.
Step 6: We produce the following contingency table.

Row-Column	O	E	$O-E$	$(O-E)^2$	$\frac{(O-E)^2}{E}$
$1-1$	72	45.32	26.68	711.822	15.7066
$1-2$	17	28.40	−11.40	129.960	4.5761
$1-3$	22	37.29	−15.29	233.784	6.2694
$2-1$	86	112.68	−26.68	711.822	6.3172
$2-2$	82	70.60	11.40	129.960	1.8408
$2-3$	108	92.71	15.29	233.784	2.5217
	276	276.0			37.2320

So $\chi^2 = 37.232$.
Step 7: Since χ^2 falls in the rejection region, we reject H_0. Age and the side to which the saggital crest deflects are not independent.

18. 389.

20. There are 4 degrees of freedom. DF = (rows − 1)(columns − 1). In this case, $(3-1)(3-1) = 4$.

22. Row 1, Column 3.

24. 12,002.

26. There are 4 degrees of freedom. DF = (rows − 1)(columns − 1). In this case, $(5-1)(2-1) = 4$.

28. No other information is need since the p-value = 0.0000. We will reject H_0 at any level.

30. 2.

32. $\chi^2 = 0.281$.

Chapter 12 Linear Regression and Correlation

Answers to Even-Numbered Self-Testing Review

12-1 Introductory Concepts

2. The dependent or response variable is the one that is to be estimated. An independent or explanatory variable is used to explain variations in the dependent variable.

4. Negative linear.

6. No linear.

8. No linear.

10. Negative.

12. None.

14. Negative.

16. Positive.

18.

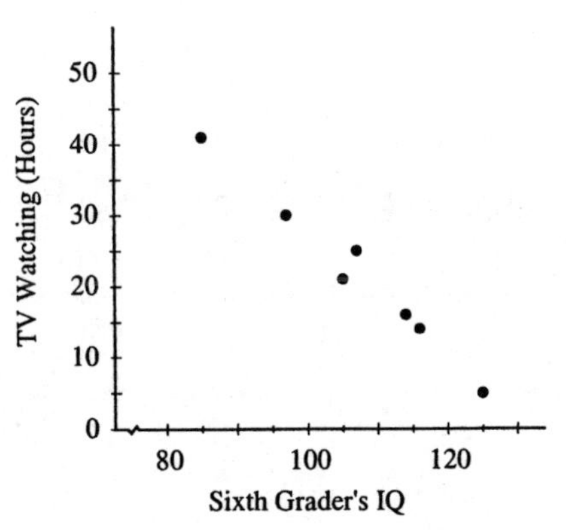

20.

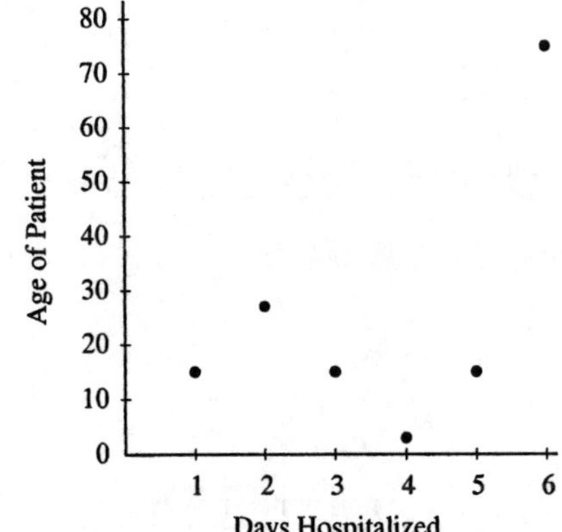

12-2 Simple Linear Regression Analysis

2. $\Sigma y = 150$; $\Sigma(y^2) = 4,042$; $(\Sigma y)^2 = 22,500$.

4. $\overline{x} = 8.125$ $\overline{y} = 18.75$.

6. $\widehat{y} = -3.3889 + 2.7248x$.

8. $\widehat{y} = -3.3889 + 2.7248(3) = 4.79$ feet.

10. $\Sigma x = 749;\ \Sigma(x^2) = 81,185;\ (\Sigma x)^2 = 561,001.$

12. $\Sigma xy = 15348.$

14. $b = -0.879;\ a = 115.776.$

16. $\hat{y} = 115.776 - 0.879(91) = 35.787$ hours.

18. $s_{y.x} = \sqrt{\dfrac{18.19}{5}} = 1.907.$

20. $\Sigma y = 280;\ \Sigma(y^2) = 10,000;\ (\Sigma y)^2 = 78,400.$

22. $\overline{x} = 15.1111;\ \overline{y} = 31.1111.$

24. $\hat{y} = 4.6154 + 1.7534x.$

26. $\hat{y} = 4.6154 + 1.7534(5) = 13.38$ cups.

28. $b = 1.68;\ a = 0.02.$

30. $\hat{y} = 0.02 + 1.68(15.0) = 25.22$ min.

32. $s_{y.x} = \sqrt{\dfrac{1.0390}{15}} = 0.2632.$

34. $\hat{y} = -90 + 1.08x.$

36. $\hat{y} = -90 + 1.08(6,500) = 6,930$ Hz.

12-3 Relationship Tests and Prediction Intervals in Simple Linear Regression Analysis

2. $s_{y.x} = \sqrt{\dfrac{155.75}{6}} = 5.095.$

4. For the *average* number of office visits, the range is:
$8.13 \pm 2.447(5.095)\sqrt{\dfrac{1}{8} + \dfrac{(5-9.375)^2}{99.875}} = 8.13 \pm 2.447(5.095)\sqrt{0.3116} = 8.13 \pm 7.0156$ which is 1.1144 *to* 15.1456 visits.

6. $\hat{y} = 20.06 + 6.811x.$

8. Step 1. H_0: $B = 0$ and H_1: $B \neq 0$.
Step 2. $\alpha = 0.05$.
Step 3. We use the t distribution.
Step 4. Since we have $21 - 2 = 19$ degrees of freedom (df), the t value in the 0.025 column and 19 df row is 2.093.
Step 5. We'll reject H_0 in favor of H_1 if $t < -2.093$ or if $t > 2.093$.

<u>Step 6.</u> To calculate the test statistic, we first need to obtain the estimated standard error of b:

$$s_b = \frac{s_{y.x}}{\sqrt{\Sigma(x^2) - \dfrac{(\Sigma x)^2}{n}}} = \frac{14.38}{\sqrt{20.9523}} \approx 3.140.$$ Then the test statistic is $t = \frac{b}{s_b} = \frac{6.811}{3.140} \approx 2.17$.

<u>Step 7.</u> Since 2.17 is greater than 2.093, we reject the H_0. A meaningful relationship does exist between the variables.

10. For the *specific* behavior score value, the interval is:

$$40.50 \pm 2.093(14.3751)\sqrt{1 + \frac{1}{21} + \frac{(3 - 3.6191)^2}{20.9524}} = 40.50 \pm 2.093(14.3751)\sqrt{1.0659} =$$

40.50 ± 31.06 or 9.44 to 71.56.

12. $s_{y.x} = \sqrt{\dfrac{14,315}{8}} = 42.30.$

14. For the *average* gross income, the range is:

$$65.64 \pm 2.306(42.30)\sqrt{\frac{1}{10} + \frac{(35 - 30.2)^2}{1621.6}} = 65.64 \pm 2.306(42.30)\sqrt{0.1142} = 65.64 \pm 32.96$$ which is 32.68 to 98.60.

16. $\hat{y} = 6.1846 - 0.0000016x.$

18. <u>Step 1.</u> H_0: $B = 0$ and H_1: $B \neq 0$.

<u>Step 2.</u> $\alpha = 0.05$.

<u>Step 3.</u> We use the t distribution.

<u>Step 4.</u> Since we have $13 - 2 = 11$ degrees of freedom (df), the t value is 2.201.

<u>Step 5.</u> We'll reject H_0 in favor of H_1 if $t < -2.201$ or if $t > 2.201$.

<u>Step 6.</u> To calculate the test statistic, we first need to obtain the estimated standard error of b:

$$s_b = \frac{s_{y.x}}{\sqrt{\Sigma(x^2) - \dfrac{(\Sigma x)^2}{n}}} = \frac{0.3926}{\sqrt{3414987 - \dfrac{(6053)^2}{13}}} = \frac{0.3926}{\sqrt{596617.1}} \approx 0.0005083.$$ Then the test statistic is

$$t = \frac{b}{s_b} = \frac{-0.0000016}{0.0005083} \approx -0.00315.$$

<u>Step 7.</u> Since t does not fall in the rejection region, we fail to reject H_0. A meaningful relationship does not exist between the variables.

20. For the *specific* serum Cholesterol level when Calcium is 500, the interval is:

$$6.184 \pm 2.201(0.3926)\sqrt{1 + \frac{1}{13} + \frac{(500 - 465.3)^2}{3414983.72}} = 6.184 \pm 0.8969$$ or 5.2871 to 7.0809.

22. $s_{y.x} = \sqrt{\dfrac{383672}{13}} = 171.8.$

24. For the *average* thickness of SiO_2 film, the interval is:

$$1269.7 \pm 3.012(171.8)\sqrt{\frac{1}{15} + \frac{(25 - 24.9333)^2}{1148.9333}} = 1269.7 \pm 3.012(171.8)\sqrt{0.06667} = 1269.7 \pm 133.6$$

which is 1136.1 to 1403.3.

26. The slope is 1.2089 and the y-intercept is 25.687.

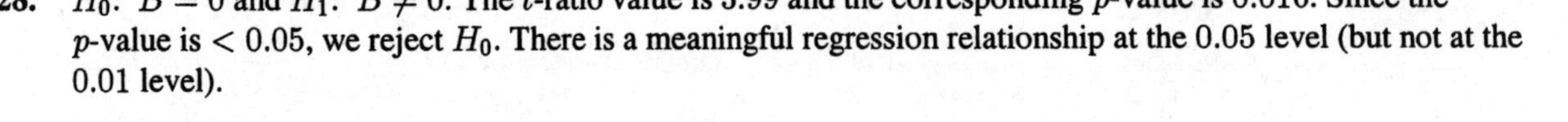

28. H_0: $B = 0$ and H_1: $B \neq 0$. The t-ratio value is 3.99 and the corresponding p-value is 0.016. Since the p-value is < 0.05, we reject H_0. There is a meaningful regression relationship at the 0.05 level (but not at the 0.01 level).

30. The slope is -0.9078 and the y-intercept is 2.4682.

32. The slope is 0.96055 and the y-intercept is 1.812.

12-4 Simple Linear Correlation Analysis

2. The explained variation $\Sigma(\widehat{y} - \overline{y})^2 = 6248.1$.

4. $r^2 = \dfrac{7810}{6248.1} = 0.80$.

6. $r = +\sqrt{0.80} = 0.894$ (r is positive since the slope is positive).

8. The explained variation $\Sigma(\widehat{y} - \overline{y})^2 = 4830.4$.

10. $r^2 = \dfrac{7184.9}{4830.4} = 0.672$.

12. $r = +\sqrt{0.672} = 0.820$ (r is positive since the slope is positive).

14. The explained variation $\Sigma(\widehat{y} - \overline{y})^2 = 26.106$.

16. $r^2 = \dfrac{28}{26.106} = 0.932$.

18. $r = -\sqrt{0.932} = -0.965$ (r is negative since the slope is negative).

20. $\text{SSR} = 129141472$.

22. $r^2 = 0.171$.

24. $\text{SST} = 5.8943$.

26. $\text{SSE} = 5.8884$.

28. $r = 0.032$.

30. $\text{SSR} = 681844$.

32. $r^2 = 0.949$.

12-5 Multiple Linear Regression and Correlation

2. YEARS and POSTHSED are the two independent variables.

4. SALARY95 $= \$29,436$. The constant in a regression equation is the value of the dependent variable when all the independent variables are zero.

6. When the number of years of experience is held constant and the number of post high school years of education is increased by 1, then the estimate of the average salary will be increased by $\$1,306$.

8. $s_{y.x_1x_2} = \sqrt{\dfrac{1912102400}{191}} = 3164.$

10. H_0: $B_{\text{POSTHSED}} = 0$ and H_1: $B_{\text{POSTHSED}} \neq 0$. The p-value for this coefficient is 0.0000, and is < 0.05, so we reject the H_0. The relationship is significant.

12. $r^2 = 0.886$.

14. Use the given formula, substituting 0 for false and 1 for true:

$$\begin{aligned}\text{Salary} &= \$22,044 + \$23,649(0) + \$15,766(0) + \$7,883(1) + \$1,152(8) \\ &\quad + \$8,392(0) + \$8,884(1) + \$3,968(0) + \$7,976(1) - \$8392(0) \\ &= \$56,003.\end{aligned}$$

16. H_0: $B_{\text{LO}} = B_{\text{THD}} = B_{\text{PF}} = B_{\text{Lamp}} = B_{\text{Blst}} = B_{\text{LiteType}} = 0$, H_1: At least one of B_{LO} or B_{THD} or B_{PF} or B_{Lamp} or B_{Blst} or $B_{\text{LiteType}} \neq 0$. The F statistic is 34.66, with a p-value of 0.007. Since the p-value < 0.05, we reject H_0 and conclude that at least one of the predictors are useful in this model.

18. H_0: $B_{\text{LiteType}} = 0$, H_1: $B_{\text{LiteType}} \neq 0$. The corresponding t-ratio is 3.56, with a p-value of 0.038. Since p-value < 0.05, we reject H_0. This predictor is significant in the model when all the others are present.

20. H_0: $B_{92} = B_{93} = B_{94} = B_{95} = 0$, H_1: At least one of B_{92} or B_{93} or B_{94} or $B_{95} \neq 0$. The F statistic is 46.54, with a p-value of 0.000. Since the p-value < 0.01, we reject H_0 and conclude that at least one of the predictors are useful in this model.

22. H_0: $B_{94} = 0$, H_1: $B_{94} \neq 0$. The corresponding t-ratio is -1.97, with a p-value of 0.055. Since p-value > 0.05, we fail to reject H_0. This predictor is not significant in the model when all the others are present.

24. H_0: $B_{\text{LT}} = B_{\text{MaxH}} = 0$, H_1: At least one of B_{LT} or $B_{\text{MaxH}} \neq 0$. The F statistic is 6.08, with a p-value of 0.021. Since the p-value < 0.05, we reject H_0 and conclude that at least one of the predictors are useful in this model.

26. H_0: $B_{\text{MaxH}} = 0$, H_1: $B_{\text{MaxH}} \neq 0$. The corresponding t-ratio is 0.63, with a p-value of 0.543. Since p-value > 0.05, we fail to reject H_0. This predictor does not appear useful in the model when the other predictor is also present in the model.

28. $a = 5.17$, $b_1 = 1.61$, and $b_2 = 0.0126$.

30. H_0: $B_{\text{LT}} = 0$, H_1: $B_{\text{LT}} \neq 0$. The corresponding t-ratio is 2.13, with a p-value of 0.062. Since p-value > 0.05, we fail to reject H_0. This predictor does not appear useful in the model when the other predictor is also present in the model.

32. $r^2 = 36.2\%$.

34. H_0: $B_{\text{Age}} = 0$, H_1: $B_{\text{Age}} \neq 0$. The corresponding t-ratio is -0.18, with a p-value of 0.868. Since p-value > 0.01, we fail to reject H_0. This predictor does not appear useful in the model when the other predictor is also present in the model.

36. H_0: $B_{\text{SolarRad}} = B_{\text{SoilTemp}} = B_{\text{VapPress}} = B_{\text{EndSpeed}} = B_{\text{RelHumid}} = B_{\text{DewPoint}} = B_{\text{AirTemp}} = 0$, H_1: At least one of B_{SolarRad} or B_{SoilTemp} or B_{VapPress} or B_{EndSpeed} or B_{RelHumid} or B_{DewPoint} or $B_{\text{AirTemp}} \neq 0$. The F statistic is 4.54, with a p-value of 0.001. Since the p-value < 0.01, we reject H_0 and conclude that at least one of the predictors are useful in this model.

38. H_0: $B_{\text{RelHumid}} = 0$, H_1: $B_{\text{RelHumid}} \neq 0$. The corresponding t-ratio is 0.82, with a p-value of 0.416. Since p-value > 0.01, we fail to reject H_0. This predictor does not appear useful in the model when the other predictor is also present in the model.

40. *Volume* $= -183 + 2.92($*Barkthck*$) + 15.0($*Diam16*$) + 1.55($*DiamBrst*$) - 0.0472($*FrmClass*$)$
$+ 0.0770($*Height*$)$.

42. H_0: $B_{\text{Diam16}} = 0$, H_1: $B_{\text{Diam16}} \neq 0$. The corresponding t-ratio is 8.65, with a p-value of 0.000. Since p-value < 0.01, we reject H_0. This predictor is significant in the model when all the others are present.

Answers to Even-Numbered Exercises

2. We compute the linear regression equation using the table below.

Car	Luggage Capacity (x)	Gas Consumption (y)	xy	x^2	y^2
1	11	28	308	121	784
2	11	29	319	121	841
3	13	27	351	169	729
4	14	25	350	196	625
5	14	23	322	196	529
6	14	24	336	196	576
7	15	22	330	225	484
8	19	20	380	361	400
9	14	24	336	196	576
10	11	30	330	121	900
11	20	18	360	400	324
12	20	19	380	400	361
TOTAL	176	289	4102	2702	7129

$$\overline{x} = \frac{\Sigma x}{n} = \frac{176}{12} = 14.6667.$$

$$\overline{y} = \frac{\Sigma y}{n} = \frac{289}{12} = 24.0833.$$

$$b = \frac{n(\Sigma xy) - (\Sigma x)(\Sigma y)}{n(\Sigma x^2) - (\Sigma x)^2} = \frac{12(4102) - (176)(289)}{12(2702) - (176)^2}$$

$$= \frac{-1640}{1448} = -1.1326.$$

$$a = \overline{y} - b\overline{x} = 24.0833 - (-1.1326)(14.6667)$$

$$= 40.695.$$

Thus the regression equation is:

$\widehat{y} = 40.695 - 1.1326x.$

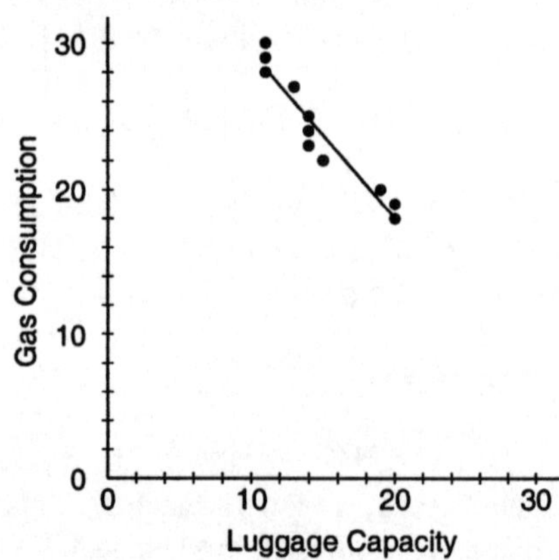

Using the Regress command in MINITAB we get:

```
The regression equation is
Gas = 40.7 - 1.13 Luggage

Predictor          Coef        StDev          T         P
Constant         40.695        1.624      25.06     0.000
Luggage         -1.1326       0.1082     -10.47     0.000

S = 1.189         R-Sq = 91.6%      R-Sq(adj) = 90.8%

Analysis of Variance

Source            DF          SS          MS          F         P
Regression         1      154.79      154.79     109.56     0.000
Residual Error    10       14.13        1.41
Total             11      168.92
```

4. We compute the linear regression equation using the worksheet below.

Car	Luggage Capacity (x)	Gas Consumption (y)	$\widehat{y}$	$(y-\widehat{y})$	$(y-\widehat{y})^2$
1	11	28	28.2414	−0.24140	0.05827
2	11	29	28.2414	0.75860	0.57547
3	13	27	25.9762	1.02380	1.04817
4	14	25	24.8436	0.15640	0.02446
5	14	23	24.8436	−1.84360	3.39887
6	14	24	24.8436	−0.84360	0.71166
7	15	22	23.7110	−1.71100	2.92753
8	19	20	19.1806	0.81940	0.67141
9	14	24	24.8436	−0.84360	0.71166
10	11	30	28.2414	1.75860	3.09267
11	20	18	18.0480	−0.04800	0.00230
12	20	19	18.0480	0.95200	0.90630
TOTAL	176	289	289.0624	−0.06240	14.12877

$$s_{y.x} = \sqrt{\frac{\Sigma(y-\widehat{y})^2}{n-2}} = \sqrt{\frac{14.12877}{12-2}} = 1.1886 \text{ mpg.}$$

6. Using Formula 12.8, with $t_{\alpha/2} = 2.228$, $s_{y.x} = 1.1886$, and $x_g = 12$, we get the interval:

$$\widehat{y} \pm t_{\alpha/2}\left[s_{y.x}\sqrt{\frac{1}{n} + \frac{(x_g - \overline{x})^2}{\Sigma(x^2) - \frac{(\Sigma x)^2}{n}}}\right] = 27.11 \pm (2.228)\cdot(1.1886)\cdot\sqrt{\frac{1}{12} + \frac{(12-14.6667)^2}{2702 - \frac{(176)^2}{12}}}$$

$$= 27.11 \pm (2.228)\cdot(1.1886)\cdot(0.3772) = 27.11 \pm 0.999.$$

Thus the interval is 26.111 to 28.109 mpg.

Using the command Regress and the subcommand Predict in MINITAB, we get the following lines added to the MINITAB output shown in Exercise 2, the desired interval is underlined.

Note: Many output lines have been omitted.

```
Predicted Values

     Fit    StDev Fit         95.0% CI                 95.0% PI
  27.104        0.448    ( 26.105,   28.103)    ( 24.273,   29.934)
```

8. Using Formula 12.11, $r^2 = \dfrac{a(\Sigma y) + b(\Sigma xy) - n(\overline{y})^2}{\Sigma(y^2) - n(\overline{y})^2}$, we get

$$r^2 = \frac{(40.695)(289) + (-1.1326)(4102) - 12(24.0833)^2}{7129 - 12(24.0833)^2} = 0.9167.$$

10.

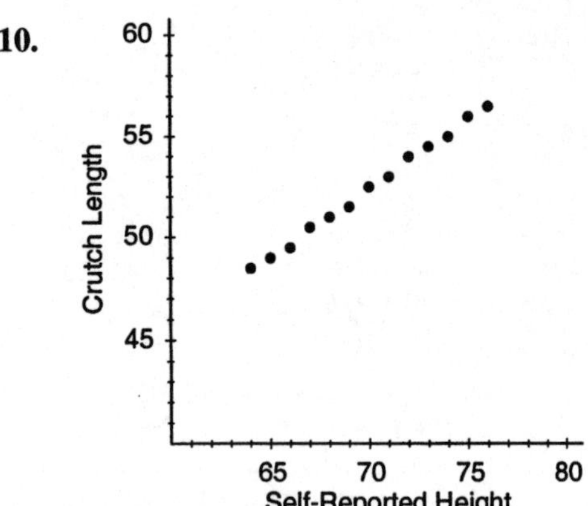

12. Using the regression equation we found in Exercise 11, $\widehat{y} = 4.5382 + 0.68407(70.5) = 52.77$ inches.

14. Step 1. H_0: $B = 0$ and H_1: $B \neq 0$.
Step 2. $\alpha = 0.05$.
Step 3. t distribution.
Step 4. With 11 df the critical t value corresponding to $t_{.025}$ is +2.201.
Step 5. Reject H_0 and accept H_1 if the test statistic, $t < -2.201$ or if $t > +2.201$.
Step 6. Using the values from the table we computed in Exercise 2 to find the regression equation we first find $s_b = \dfrac{s_{y.x}}{\sqrt{\Sigma(x^2) - \dfrac{(\Sigma x)^2}{n}}} = \dfrac{0.1532}{\sqrt{63882 - \dfrac{(910)^2}{13}}} = \dfrac{0.1532}{\sqrt{182}} = 0.0114.$

Thus $t = \dfrac{b - B_{H_0}}{s_b} = \dfrac{0.68407}{0.0114} = 60.006$. (MINITAB rounds values less than we do and reports this test statistic value as 60.39.)

Step 7. Since $60.006 > 2.201$, we reject H_0. There is a meaningful regression relationship--that is, the relationship is significant--between self-reported height and crutch length.

16. Using Formula 12.09, with the values $\widehat{y} = 51.05496$, $t_{\alpha/2} = 2.201$, and $s_{y.x} = 0.1432$.

$$\widehat{y} \pm t_{\alpha/2}\left[s_{y.x}\sqrt{1 + \frac{1}{n} + \frac{(x_g - \overline{x})^2}{\Sigma(x^2) - \dfrac{(\Sigma x)^2}{n}}}\right]$$

$$= 51.05496 \pm (2.201)\cdot(0.1532)\cdot\sqrt{1 + \frac{1}{13} + \frac{(68-70)^2}{63,882 - \dfrac{(910)^2}{13}}}$$

$$= 51.05496 + 2.201 \cdot 0.1532 \cdot \sqrt{1.0989} = 51.05496 + 0.35347.$$

Thus we get the interval 50.7015 to 51.4084 inches.

Using the command Regress with the subcommand Predict in MINITAB, we get 95% P.I. shown underlined on the last line of the printout.

```
Predicted Values

     Fit   StDev Fit         95.0% CI               95.0% PI
 51.0549     0.0481    ( 50.9492, 51.1607)   ( 50.7024, 51.4075)
```

18. Using Formula 12.11 to first find the coefficient of determination we get:

$$r^2 = \frac{a(\Sigma y) + b(\Sigma xy) - n(\overline{y})^2}{\Sigma(y^2) - n(\overline{y})^2} = \frac{(4.5382)(681.5) + (0.68407)(47,829.5) - 13(52.4231)^2}{35,811.75 - 13(52.4231)^2} = 0.997.$$

Thus the coefficient of correlation, $r = \sqrt{0.997} = 0.9985$.

20. Step 1. H_0: $B = 0$ and H_1: $B \neq 0$.

Step 2. $\alpha = 0.05$.

Step 3. t distribution.

Step 4. With 20 df the critical t value corresponding to $t_{.025}$ is 2.086.

Step 5. Reject H_0 and accept H_1 if the test statistic $t < -2.086$ or if $t > +2.086$.

Step 6. Using the values from the table we computed in Exercise 19 to find the regression equation we first find

$$s_{y.x} = \sqrt{\frac{\Sigma(y^2) - a(\Sigma y) - b(\Sigma xy)}{n-2}} = \sqrt{\frac{1869.98 - 2.410(198.8) - 0.7453(1829.7)}{20}} = 1.166.$$

$$s_b = \frac{s_{y.x}}{\sqrt{\Sigma(x^2) - \frac{(\Sigma x)^2}{n}}} = \frac{1.169}{\sqrt{1,822.5 - \frac{(195.6)^2}{22}}} = \frac{1.166}{\sqrt{83.438}} = 0.1276. \text{ Thus } t = \frac{b}{s_b} = \frac{0.7453}{0.1276} = 5.840.$$

Step 7. Since $5.84 > 2.086$, we reject H_0. There is a meaningful regression relationship.

22. Using Formula 12.09, with the values $\hat{y} = 6.882$, $t_{\alpha/2} = 2.086$, and $s_{y.x} = 1.169$.

$$\hat{y} \pm t_{\alpha/2}\left[s_{y.x}\sqrt{1 + \frac{1}{n} + \frac{(x_g - \overline{x})^2}{\Sigma(x^2) - \frac{(\Sigma x)^2}{n}}}\right] = 6.882 \pm (2.086)\cdot(1.166)\cdot\sqrt{1 + \frac{1}{22} + \frac{(12 - 8.891)^2}{1822.5 - \frac{(195.6)^2}{22}}}$$

$$= 6.882 \pm (2.086)\cdot(1.166)\cdot\sqrt{1.1641} = 6.882 \pm 2.624 \text{ or } 4.258 \text{ to } 9.506.$$

Using MINITAB we get:

```
Predicted Values

    Fit  StDev Fit          95.0% CI               95.0% PI
  6.882      0.445    (   5.954,    7.810)  (   4.279,    9.484)
```

24. We compute the linear regression equation using the table below.

Student	PIAT (x)	WRAT (y)	xy	x^2	y^2
1	17	20	340	289	400
2	21	28	588	441	784
3	22	32	704	484	1024
4	25	33	825	625	1089
5	25	38	950	625	1444
6	31	40	1240	961	1600
7	31	44	1364	961	1936
8	35	45	1575	1225	2025
9	36	46	1656	1296	2116
10	44	53	2332	1936	2809
TOTAL	287	379	11574	8843	15227

$$\overline{x} = \frac{\Sigma x}{n} = \frac{287}{10} = 28.7. \qquad \overline{y} = \frac{\Sigma y}{n} = \frac{379}{10} = 37.9.$$

$$b = \frac{n(\Sigma xy) - (\Sigma x)(\Sigma y)}{n(\Sigma x^2) - (\Sigma x)^2} = \frac{10(11,574) - (287)(379)}{10(8,843) - (287)^2} = \frac{6,967}{6,061} = 1.15.$$

$a = \overline{y} - b\overline{x} = 37.9 - (1.1495)(28.7) = 4.91.$

Thus the regression equation is: $\hat{y} = 4.91 + 1.15x$.

Using the regression command in MINITAB we get:

```
Regression Analysis

The regression equation is
WRAT = 4.91 + 1.15 PIAT

Predictor        Coef      StDev         T        P
Constant        4.910      3.364      1.46    0.183
PIAT           1.1495     0.1131     10.16    0.000

S = 2.785       R-Sq = 92.8%      R-Sq(adj) = 91.9%

Analysis of Variance

Source            DF        SS          MS          F        P
Regression         1    800.84      800.84     103.24    0.000
Residual Error     8     62.06        7.76
Total              9    862.90
```

26. We compute the standard error of the estimate using the table below.

Student	PIAT (x)	WRAT (y)	$\hat{y}$	$(y-\hat{y})$	$(y-\hat{y})^2$
1	17	20	24.4515	−4.45150	19.8159
2	21	28	29.0495	−1.04950	1.1014
3	22	32	30.1990	1.80100	3.2436
4	25	33	33.6475	−0.64750	0.4193
5	25	38	33.6475	4.35250	18.9443
6	31	40	40.5445	−0.54450	0.2965
7	31	44	40.5445	3.45550	11.9405
8	35	45	45.1425	−0.14250	0.0203
9	36	46	46.2920	−0.29200	0.0853
10	44	53	55.4880	−2.48800	6.1901
TOTAL	287	379	379.0065	−0.00650	62.0572

$$s_{y.x} = \sqrt{\frac{\Sigma(y-\hat{y})^2}{n-2}} = \sqrt{\frac{62.0572}{10-2}} = 2.785.$$

28. We compute the linear regression equation using the table below.

Employee	Baseline BMI (x)	Two-Year BMI (y)	xy	x^2	y^2
1	26.97	26.02	701.7594	727.3809	677.0404
2	25.64	25.87	663.3068	657.4096	669.2569
3	25.12	25.02	628.5024	631.0144	626.0004
4	25.57	25.46	651.0122	653.8249	648.2116
5	26.09	25.70	670.5130	680.6881	660.4900
6	26.17	26.10	683.0370	684.8689	681.2100
7	25.92	26.24	680.1408	671.8464	688.5376
8	25.68	26.57	682.3176	659.4624	705.9649
9	25.07	24.57	615.9699	628.5049	603.6849
10	25.70	25.18	647.1260	660.4900	634.0324
11	26.61	26.84	714.2124	708.0921	720.3856
12	26.34	26.31	693.0054	693.7956	692.2161
13	26.34	26.22	690.6348	693.7956	687.4884
14	25.70	25.61	658.1770	660.4900	655.8721
15	26.30	26.42	694.8460	691.6900	698.0164
16	25.84	25.16	650.1344	667.7056	633.0256
TOTAL	415.06	413.29	10724.6951	10771.0594	10681.4325

$$\overline{x} = \frac{\Sigma x}{n} = \frac{415.06}{16} = 25.941. \qquad \overline{y} = \frac{\Sigma y}{n} = \frac{413.29}{16} = 25.831.$$

$$b = \frac{n(\Sigma xy) - (\Sigma x)(\Sigma y)}{n(\Sigma x^2) - (\Sigma x)^2} = \frac{16(10{,}724.6951) - (415.06)(413.29)}{16(10{,}771.0594) - (415.06)^2} = 0.8846.$$

$a = \overline{y} - b\overline{x} = 25.831 - (0.8846)(25.941) = 2.8836.$

Thus the regression equation is: $\widehat{y} = 2.884 + 0.8846x$.

Using the regression command in MINITAB we get:

```
Regression Analysis

The regression equation is
Two-Year = 2.88 + 0.885 Baseline

Predictor          Coef        StDev            T        P
Constant          2.883        5.945         0.48    0.635
Baseline         0.8846       0.2291         3.86    0.002

S = 0.4516        R-Sq = 51.6%       R-Sq(adj) = 48.1%

Analysis of Variance

Source              DF          SS           MS          F         P
Regression           1      3.0393       3.0393      14.90     0.002
Residual Error      14      2.8550       0.2039
Total               15      5.8943
```

30. We compute the standard error of the estimate using the table below.

Employee	Baseline BMI (x)	Two-Year BMI (y)	$\widehat{y}$	$(y - \widehat{y})$	$(y - \widehat{y})^2$
1	26.97	26.02	26.7484	−0.728449	0.530638
2	25.64	25.87	25.5714	0.298601	0.089163
3	25.12	25.02	25.1112	−0.091200	0.008317
4	25.57	25.46	25.5094	−0.049451	0.002445
5	26.09	25.70	25.9697	−0.269649	0.072711
6	26.17	26.10	26.0405	0.059550	0.003546
7	25.92	26.24	25.8192	0.420800	0.177072
8	25.68	26.57	25.6068	0.963199	0.927753
9	25.07	24.57	25.0669	−0.496950	0.246959
10	25.70	25.18	25.6245	−0.444500	0.197581
11	26.61	26.84	26.4299	0.410150	0.168223
12	26.34	26.31	26.1909	0.119099	0.014185
13	26.34	26.22	26.1909	0.029099	0.000847
14	25.70	25.61	25.6245	−0.014500	0.000210
15	26.30	26.42	26.1555	0.264501	0.069961
16	25.84	25.16	25.7484	−0.588400	0.346215
	415.06	413.29	413.41	−0.11810	2.8558

$$s_{y.x} = \sqrt{\frac{\Sigma(y - \widehat{y})^2}{n-2}} = \sqrt{\frac{2.8558}{16-2}} = 0.4516.$$

This value is also shown on the MINITAB output in Exercise 28.

32. Using Formula 12.11 to first find the coefficient of determination we get:

$$r^2 = \frac{a(\Sigma y) + b(\Sigma xy) - n(\overline{y})^2}{\Sigma(y^2) - n(\overline{y})^2}$$

$$= \frac{(2.8836)(413.29) + (0.8846)(10{,}724.6951) - 16(25.831)^2}{10{,}681.4325 - 16(25.831)^2} = 0.5336.$$

Thus the coefficient of correlation, $r = \sqrt{0.5336} = 0.7305$. (MINITAB reports $r^2 = 0.516$.)

34. Using the regression equation we found in Exercise 35, $\widehat{y} = -11.50 + 0.633(52.00) = \$21.42$.

36. Using Formula 12.11 the coefficient of determination is:

$$r^2 = \frac{a(\Sigma y) + b(\Sigma xy) - n(\overline{y})^2}{\Sigma(y^2) - n(\overline{y})^2} = \frac{(-11.49)(125.375) + (0.6327)(6,420.30) - 6(20.896)^2}{2,621.86 - 6(20.896)^2}$$
$$= 0.8527.$$

Using different rounding, MINITAB reports $r^2 = 0.824$.

38. Using the regression equation we found in Exercise 37, $\widehat{y} = -2.587 + 0.7933(12) = 6.933$ datagrams lost.

40. Using the point estimate from Exercise 38, $\widehat{y} = 6.93$ when $x_g = 12$. Then with $t_{\alpha/2} = 2.101$, $s_{y.x} = 4.326$, and $x_g = 8$, we get the interval:

$$\widehat{y} \pm t_{\alpha/2}\left[s_{y.x}\sqrt{\frac{1}{n} + \frac{(x_g - \overline{x})^2}{\Sigma(x^2) - \frac{(\Sigma x)^2}{n}}}\right] = 6.933 \pm (2.101)\cdot(4.318)\cdot\sqrt{\frac{1}{20} + \frac{(12 - 12.4)^2}{4,032 - \frac{(248)^2}{20}}}$$
$$= 6.933 \pm (2.101)\cdot(4.318)\cdot(0.22398)$$
$$= 6.933 \pm 2.032.$$

Thus the interval is 4.901 to 8.965 datagrams lost.

Using the command Regress and the subcommand Predict in MINITAB, we get the following lines added to the MINITAB output shown in Exercise 11, the desired interval is underlined.

Note: Many output lines have been omitted.

```
Predicted Values

   Fit   StDev Fit          95.0% CI               95.0% PI
  6.933      0.967     (   4.901,    8.965)   (  -2.365,   16.230)
```

42. Using Formula 12.11 the coefficient of determination is:

$$r^2 = \frac{a(\Sigma y) + b(\Sigma xy) - n(\overline{y})^2}{\Sigma(y^2) - n(\overline{y})^2} = \frac{(-2.587)(145) + (0.7933)(2557) - 20(7.25)^2}{1989 - 20(7.25)^2}$$
$$= 0.642.$$

44. *Behavior* = 6.3 + 9.08(*developm*) + 0.119(*cbfmean*).

46. Developm score and cbfmean score.

48. 0.616

50. p value for cbfmean is 0.598. Since $0.598 > 0.05$, we fail to reject H_0. The relationship between behavior and cbfmean is not significant at the 0.05 level. (The t ratio method confirms this. Here the t ratio is 0.54, while with 18 df $t_{.025}$ is +2.101, so since 0.54 is within ±2.101, we fail to reject H_0.)

52. 14 independent variables.

```
Analysis of Variance

Source            DF        SS         MS        F        P
Regression        14      8483.6      606.0     2.27    0.017
Residual Error    51     13604.9      226.8
Total             65     22088.5
```

54. The y-intercept would corresponds to all 14 variables are 0 and that can not happen.

56. H_0: That all the B's $= 0$, H_1: At least one of B's $\neq 0$. The F statistic is 2.27, with a p-value of 0.017. Since the p-value < 0.05, we reject H_0 and conclude that at least one of the predictors are useful in this model.

58. H_0: $B_{TDS} = 0$, H_1: $B_{TDS} \neq 0$. The corresponding t-ratio is -2.05, with a p-value of 0.045. Since p-value < 0.05, we reject H_0. This predictor is significant in the model when all the others are present.

60.

Variable	Coefficient
Constant	1796.3
pH	−142.2
Conducty	0.2414
COD	2.5980

62. H_0: $B_{pH} = 0$, H_1: $B_{pH} \neq 0$. The corresponding t-ratio is -1.07, with a p-value of 0.292. Since p-value > 0.05, we fail to reject H_0. This predictor is not significant in the model when all the others are present.

Chapter 13 Nonparametric Statistical Methods

Solutions to Even-Numbered Self-Testing Review

13-1 Nonparametric Methods: uses, Benefits, and Limitions

2. You use a nonparametric method if the sample size is small and no assumption about the shape of the population distribution can be made. Nonparametric methods are also appropriate for testing ordinal (rank) and nominal data.

4. If nominal or ordinal data are the only facts available, nonparametric methods can be used while parametric methods aren't available to an analyst. If sample sizes are small, nonparametric methods can be used. Little or no information about the population(s) being sampled is required to use nonparametric procedures, and these procedures are easier to understand and to use than parametric methods. Calculation is usually easier with nonparametric methods.

13-2 Sign Tests

2. The null hypothesis states that $p = 0.5$, meaning that 50 percent of the signs are positive. That is, a positive sign and a negative sign are equally likely to occur.

4. **(a)** Since $n =$ the total number of positive and negative scores, $n = 15 + 5 = 20$ (zeros and ties are ignored). There are 5 negative scores (which is less than the 15 positive scores) so $r = 5$.
   (b) The sum of the binomial probabilities is 0.0207. At the 0.05 level, we reject the H_0 since $0.0207 < 0.05$.

6. Step 1. H_0: $p = 0.5$ and H_1: $p > 0.5$.
   Step 2. $\alpha = 0.05$.
   Step 3. Subtracting the 2 year BMI from the baseline BMI and noting only the sign of the difference, the sign results in order are
   $+ - - + 0 - - + - - + - - + - +.$
   Thus, there are 6 plus signs, 9 minus signs, and 1 zero. That is, 6 had a higher BMI, 9 had a lower BMI, and one person stayed the same. The number of minus signs, 9, is designated as r. The total of 6 plus signs and 9 minus signs is 15, so $n = 15$ (ignore the 0). Assuming the H_0 is true, the probability of a $+$ sign is 0.5.
   Step 4. We'll use the binomial table to calculate the probability that $r \leq 9$ when $n = 15$ and $p = 0.5$.
   Step 5. We'll reject H_0 in favor of H_1 if the probability computed in the next step < 0.05.
   Step 6. The probabilities for $r = 0, 1, 2, 3, 4, 5, 6, 7, 8,$ and 9 are
   $0.0000 + 0.0005 + 0.0032 + 0.0139 + 0.0417 + 0.0916 + 0.1517 + 0.1964 + 0.1964 + 0.1527 =$ 0.8491.

Step 7. Since $0.05 < 0.8491$, we fail to reject H_0. It seems that the program had no effect on the participants' BMI.

8. Step 1. H_0: $p = 0.5$ and H_1: $p > 0.5$.
Step 2. $\alpha = 0.01$.
Step 3. The signs of the differences for *female rating – male rating* are:
$+ + - + + - + - - + + - + - + + + + + -$.
Thus, there are 13 pluses and 7 minuses. The total number of signs is 20, so $n = 20$, and the number of minus signs is 7, so $r = 7$.
Step 4. We'll use the binomial table.
Step 5. We'll reject H_0 in favor of H_1 if the probability computed in the next step < 0.01.
Step 6. Assuming $p = 0.5$, the decision probability value is 0.1316.
Step 7. Since $0.01 < 0.1316$, we fail to reject H_0. Women do not generally tend to rate these values higher than men.

10. Step 1. H_0: $p = 0.5$ and H_1: $p \neq 0.5$.
Step 2. $\alpha = 0.05$.
Step 3. The signs for the *control arm – treated arm* differences are:
$+ - - + + - + + + +$.
Thus, there are 7 pluses and 3 minuses. So $n = 10$ and $r = 3$.
Step 4. We'll use the binomial table.
Step 5. We'll reject H_0 in favor of H_1 if the probability computed in the next step < 0.05.
Step 6. Since this is a two-tailed test, the decision probability value is $2(0.1719) = 0.3438$.
Step 7. Since $0.05 < 0.3438$, we fail to reject H_0. The ultrasound did not make a significant difference in relieving pain.

12. Step 1. H_0: $p = 0.5$ and H_1: $p \neq 0.5$.
Step 2. $\alpha = 0.05$.
Step 3. There are 11 pluses and 3 minuses, $n = 14$ and $r = 3$.
Step 4. We'll use the binomial table.
Step 5. We'll reject H_0 in favor of H_1 if the probability computed in the next step < 0.05.
Step 6. The decision probability value is 0.0576.
Step 7. Since $0.05 < 0.0576$, we fail to reject H_0. The scores have not changed significantly since last year.

14. Step 1. H_0: $p = 0.5$ and H_1: $p > 0.5$.
Step 2. $\alpha = 0.05$.
Step 3. Data given in problem.
Step 4. We'll use the z distribution.
Step 5. We'll reject H_0 in favor of H_1 if the test statistic $z > 1.645$. Otherwise, fail to reject H_0.
Step 6. Then $z = \dfrac{2(14) - 25}{\sqrt{25}} = 0.6$.
Step 7. Since $z = 0.6$ does not fall in the rejection region, we fail to reject H_0. Great Buys does not have lower prices.

16. Step 1. H_0: $p = 0.5$ and H_1: $p < 0.5$.
Step 2. $\alpha = 0.10$.
Step 3. The *1991 – 1992* signs tally to 3 minuses, 1 plus, so $n = 4$ and $r = 1$.
Step 4. We'll use the binomial table.
Step 5. We'll reject H_0 in favor of H_1 if the probability computed in the next step < 0.10.
Step 6. The decision probability value is $0.0625 + 0.2500 = 0.3125$.

Step 7. Since $0.10 < 0.3125$, we fail to reject H_0. There is insignificant evidence to determine that there is a decrease in the rates.

18. Step 1. H_0: $p = 0.5$ and H_1: $p < 0.5$.
Step 2. $\alpha = 0.10$.
Step 3. The *1993 – 1994* signs tally to 4 minuses, 0 plus, so $n = 4$ and $r = 0$.
Step 4. We'll use the binomial table.
Step 5. We'll reject H_0 in favor of H_1 if the probability computed in the next step < 0.10.
Step 6. The decision probability value is 0.0625.
Step 7. Since $0.10 > 0.0625$, we reject H_0. There does seem to be a decrease in the rates.

20. Step 1. H_0: $p = 0.5$ and H_1: $p \neq 0.5$.
Step 2. $\alpha = 0.05$.
Step 3. The FatLT – CarbLT signs tally to 4 minuses, 8 plus, so $n = 12$ and $r = 4$.
Step 4. We'll use the binomial table.
Step 5. We'll reject H_0 in favor of H_1 if the probability computed in the next step < 0.05.
Step 6. The decision probability value is $(0.0002 + 0.0029 + 0.0161 + 0.0537 + 0.120) \times 2 = (0.1772) \times 2 = 0.3544$.
Step 7. Since $0.05 < 0.3544$, we fail to reject H_0. There appears to be no difference in the lactate threshold between the two diets.

22. Step 1. H_0: $p = 0.5$ and H_1: $p \neq 0.5$.
Step 2. $\alpha = 0.05$.
Step 3. The FatTimeE – CarbTimeE signs tally to 8 minuses, 4 plus, so $n = 12$ and $r = 4$.
Step 4. We'll use the binomial table.
Step 5. We'll reject H_0 in favor of H_1 if the probability computed in the next step < 0.05.
Step 6. The decision probability value is $(0.0002 + 0.0029 + 0.0161 + 0.0537 + 0.1204) \times 2 = (0.1772) \times 2 = 0.3544$.
Step 7. Since $0.05 < 0.3444$, we fail to reject H_0. There appears to be no difference in the performance time between the two diets.

13-3 The Wilcoxon Signed Rank Test

2. In a Wilcoxon test, the H_0 is that there is no real difference between the paired data.

4. The T table value is 159.

6. Step 1. H_0: There's no difference in the attention span for the two groups of twins. H_1: The twins who took the vitamin supplement had a longer attention span.
Step 2. $\alpha = 0.05$.
Step 3. The rank differences irrespective of sign are 3, 8, 4, 5, 6, 7, 1, 2.
Step 4. We use the T statistic.
Step 5. Reject H_0 in favor or H_1 if the computed $\boldsymbol{T}$ value is $<$ table T value of 5.
Step 6. The positive rank sum is $3 + 8 + 5 + 6 + 7 + 1 + 2 = 32$, and the negative rank sum is 4. The computed $\boldsymbol{T}$ value is therefore 4.
Step 7. Since the computed $\boldsymbol{T}$ value of 4 is $<$ the table T value of 5, we reject H_0. The test indicates that the vitamins have the effect of lengthening the attention span.

8. Step 1. H_0: The time for a sales presentation is the same with and without the multimedia presentation. H_1: The multimedia presentation reduces sales time.
   Step 2. $\alpha = 0.05$.
   Step 3. The rank differences irrespective of sign are 6.5, 1, ignore, 4.5, 4.5, 6.5, 2, 3, 8, 9.
   Step 4. We use the T statistic.
   Step 5. Reject H_0 in favor or H_1 if the computed $\boldsymbol{T}$ value is $<$ table T value of 8.
   Step 6. The positive rank sum is $6.5 + 2 + 9 = 17.5$, and the negative rank sum is $6.5 + 1 + 4.5 + 4.5 + 3 + 8 = 27.5$. The computed $\boldsymbol{T}$ value is 17.5.
   Step 7. Since the computed $\boldsymbol{T}$ value of 17.5 is $>$ the table T value of 8, we fail to reject H_0. The multimedia presentation does not reduce the length of time needed to make a sale.

10. Step 1. H_0: The pain is the same in the arm treated with ultrasound and in the control arm. H_1: There was a difference between arms.
   Step 2. $\alpha = 0.05$.
   Step 3. See data in Step 6.
   Step 4. We use the T statistic.
   Step 5. Reject H_0 in favor or H_1 if the computed $\boldsymbol{T}$ value is $<$ table T value of 8.
   Step 6. The positive rank sum is $4 + 8 + 2 = 14$, and the sum of the negative ranks is $10 + 5.5 + 7 + 9 + 3 + 1 + 5.5 = 41$. The computed $\boldsymbol{T}$ value is therefore 14.
   Step 7. Since the computed $\boldsymbol{T}$ value of 14 is $>$ the table T valuc of 8, wc fail to reject II_0. The ultrasound treatment doesn't seem to help much.

12. Step 1. H_0: The darker color rating is the same as the original color rating. H_1: The darker color rating is higher.
   Step 2. $\alpha = 0.05$.
   Step 3. See data in Step 6.
   Step 4. We use the T statistic.
   Step 5. Reject H_0 in favor or H_1 if the computed $\boldsymbol{T}$ value is $<$ table T value of 2.
   Step 6. The positive rank sum is 11, and the sum of the negative ranks is 10, so the computed T value is therefore 10.
   Step 7. Since the computed $\boldsymbol{T}$ value of 10 is $>$ the table T value of 2, we fail to reject H_0. The change in color didn't significantly improve the ratings.

13-4 The Mann-Whitney Test

2. The t test requires that the samples are taken from normally distributed populations. The Mann-Whitney Test only requires are the population distributions have the identical shape and equal variance.

4. A U value is computed for each of the two samples. The U statistic is the lesser of the two computed values.

6. Step 1. H_0: There's no difference in the ratings between the largest companies in the Computer/Office Equipment industry and those in the telecommunications industry. H_1: There is a difference in the ratings.
   Step 2. $\alpha = 0.05$.
   Step 3. The sum of the ranks in the computer industry is 125, the sum of the ranks in the telecommunications industry is 85, $n_1 = 10$, and $n_2 = 10$.
   Step 4. We use the U statistic.
   Step 5. Reject H_0 in favor of H_1 if the computed $\boldsymbol{U}$ statistic is $<$ the table U value of 23.

Step 6. $U_1 = 10(10) + \frac{10(11)}{2} - 125 = 30$, and $U_2 = 10(10) + \frac{10(11)}{2} - 85 = 70$. Since the lesser of these values is 30, so $30 =$ the computed $\boldsymbol{U}$ statistic.
Step 7. Since the U statistic of 30 is > 23, we fail to reject H_0. There's no significant difference in the ratings of the two groups.

8. Step 1. H_0: The treatment makes no difference in the amount of pain. H_1: There is less pain in the treated arm.
Step 2. $\alpha = 0.05$.
Step 3. $R_1 = 117.5$, $R_2 = 92.5$, $n_1 = 10$, and $n_2 = 10$.
Step 4. We use the U statistic.
Step 5. Reject H_0 in favor of H_1 if the computed $\boldsymbol{U}$ statistic is $<$ the table U value of 19.
Step 6. Since this is a right-tailed test, $\boldsymbol{U} = U_1 = 10(10) + \frac{10(11)}{2} - 117.5 = 37.5$.
Step 7. Since $37.5 > 19$, we fail to reject H_0. The treatment does not make a significant difference in pain relief.

10. Step 1. H_0: Students in both groups had the same scores. H_1: Students in the phonetically based group scored higher.
Step 2. $\alpha = 0.01$.
Step 3. $R_1 = 41.50$, $R_2 = 111.50$, $n_1 = 7$, and $n_2 = 10$.
Step 4. We use the U statistic.
Step 5. Reject H_0 in favor of H_1 if the computed $\boldsymbol{U}$ value is $<$ the table U value of 11.
Step 6. $\boldsymbol{U} = U_2 = 7(10) + \frac{10(11)}{2} - 111.5 = 13.5$.
Step 7. Since $13.5 > 11$, we fail to reject H_0. There's no difference in the scores of these groups.

12. Step 1. H_0: There's no difference in satisfaction. H_1: College grads are more satisfied with their jobs.
Step 2. $\alpha = 0.05$.
Step 3. $R_1 = 154$, $R_2 = 56$, $n_1 = 12$, and $n_2 = 8$.
Step 4. We use the U statistic.
Step 5. Reject H_0 in favor of H_1 if the computed $\boldsymbol{U}$ value is $<$ the table U value of 26.
Step 6. $\boldsymbol{U} = U_2 = 12(8) + \frac{8(9)}{2} - 56 = 76$.
Step 7. Since $76 > 26$, we fail to reject H_0. College grads are not more satisfied with their jobs.

14. Step 1. H_0: The distribution of *oakness* is the same for the two treatments. H_1: The distribution of *oakness* is not the same for the two treatments.
Step 2. $\alpha = 0.05$.
Step 3. $n_1 = 21$ and $n_2 = 21$. We compute the sum of the ranks

oak12	Rank	oak24	Rank
1.00000	3.5	0.80000	2.0
1.90000	8.0	2.90000	11.0
1.00000	3.5	3.10000	12.5
3.30000	14.5	4.60000	22.0
3.60000	17.0	6.10000	32.0
1.80000	7.0	5.70000	27.5
4.10000	18.5	4.10000	18.5
4.20000	20.0	5.00000	24.0
3.10000	12.5	3.40000	16.0
6.00000	31.0	5.70000	27.5
6.70000	37.0	6.90000	38.5
6.30000	33.0	7.30000	40.0
3.30000	14.5	5.90000	30.0
5.20000	25.0	8.10000	42.0
4.90000	23.0	6.50000	34.5
5.50000	26.0	7.50000	41.0
5.80000	29.0	6.90000	38.5
6.50000	34.5	4.30000	21.0
1.10000	5.5	2.50000	9.0
1.10000	5.5	0.30000	1.0
2.60000	10.0	6.60000	36.0
	$R_1 = 378.5$		$R_2 = 524.5$

Step 4. We use the U statistic.

Step 5. Reject H_0 in favor of H_1 if the computed $\boldsymbol{U}$ value is $<$ the table U value of 127.

Step 6. $U_1 = 21(21) + \dfrac{21(22)}{2} - 378.5 = 293.5$ and $U_2 = 21(21) + \dfrac{21(21)}{2} - 524.5 = 147.5$. Since $\boldsymbol{U}$ is the lesser, $\boldsymbol{U} = 147.5$.

Step 7. Since $\boldsymbol{U} > 127$, we fail to reject H_0. The distribution of *oakness* is the same for the two treatments.

This is confirmed by MINITAB.

```
MTB > Mann-Whitney 95.0 'oak12' 'oak24';
SUBC>   Alternative 0.

Mann-Whitney Confidence Interval and Test

oak12      N =  21     Median =        3.600
oak24      N =  21     Median =        5.700
Point estimate for ETA1-ETA2 is       -1.300
95.0 Percent CI for ETA1-ETA2 is (-2.600,0.101)
W = 378.5
Test of ETA1 = ETA2  vs  ETA1 not = ETA2 is significant at 0.0682
The test is significant at 0.0681 (adjusted for ties)

Cannot reject at alpha = 0.05
```

16. Step 1. H_0: The two different microenvironment treatments produce the same number of defects. H_1: The different microenvironment treatments produce different number of defects.

Step 2. $\alpha = 0.05$.

Step 3. $n_1 = 10$ and $n_2 = 5$. We compute the sum of the ranks.

Standard	Rank	Megasonic	Rank
53	2	26	1.0
193	11	90	4.5
113	6	546	12.0
640	13	90	4.5
800	15	120	7.0
140	9		
85	3		
658	14		
140	9		
140	9		
	$R_1 = 91$		$R_2 = 29.0$

Step 4. We use the U statistic.

Step 5. Reject H_0 in favor of H_1 if the computed $\boldsymbol{U}$ value is < the table U value of 8.

Step 6. $U_1 = 10(5) + \dfrac{10(11)}{2} - 91 = 14$ and $U_2 = 10(5) + \dfrac{5(6)}{2} - 29 = 36$. Since $\boldsymbol{U}$ is the lesser, $\boldsymbol{U} = 14$.

Step 7. Since $\boldsymbol{U} > 8$, we fail to reject H_0. The two different microenvironment treatments produce the same number of defects.

This is confirmed by MINITAB.

```
MTB > Mann-Whitney 95.0 'standard' 'megasonic';
SUBC>   Alternative 0.

Mann-Whitney Confidence Interval and Test

standard   N =  10     Median =        140.0
megasoni   N =  5      Median =         90.0
Point estimate for ETA1-ETA2 is         50.0
95.7 Percent CI for ETA1-ETA2 is (-37.1,550.1)
W = 91.0
Test of ETA1 = ETA2  vs  ETA1 not = ETA2 is significant at 0.1984
The test is significant at 0.1964 (adjusted for ties)

Cannot reject at alpha = 0.05
```

13-5 The Kruskal-Wallis Test

2. Step 1. H_0: The mean number of products sold for all three height displays is equal. H_1: At least one of the means is different.

   Step 2. $\alpha = 0.01$.

   Step 3. $N = 15$, $k = 3$, $R_1 = 51$, $R_2 = 43$, $R_3 = 26$, $n_1 = 5$, $n_2 = 5$, and $n_3 = 5$.

   Step 4. We use the χ^2 distribution.

   Step 5. Reject H_0 in favor of H_1 if the computed $\boldsymbol{H}$ value is > 9.21. Otherwise, fail to reject H_0.

   Step 6. $\boldsymbol{H} = \dfrac{12}{15(16)}\left(\dfrac{51^2}{5} + \dfrac{43^2}{5} + \dfrac{26^2}{5}\right) - 3(16) = 3.26$.

   Step 7. Since 3.26 < 9.21 , we fail to reject H_0. There's no significant difference in the number of products sold at different display heights.

4. Step 1. H_0: The mean daily traffic counts for all three locations are equal. H_1: At least one of the means is different.

   Step 2. $\alpha = 0.05$.

   Step 3. $N = 21$, $k = 3$, $R_1 = 81$, $R_2 = 68$, $R_3 = 146.5$, $n_1 = 7$, $n_2 = 7$, and $n_3 = 7$.

Step 4. We use the χ^2 distribution.
Step 5. Reject H_0 in favor of H_1 if the computed $\boldsymbol{H}$ value is > 5.99. Otherwise, fail to reject H_0.
Step 6. $\boldsymbol{H} = 0.45$.
Step 7. Since $0.45 < 5.99$, we fail to reject H_0. The traffic counts are equal for the three locations.

6. Step 1. H_0: The mean number of errors made by those from different schools is equal. H_1: At least one of the means is different.
Step 2. $\alpha = 0.01$.
Step 3. $N = 28$, $k = 4$, $R_1 = 86$, $R_2 = 87$, $R_3 = 146.5$, $R_4 = 86.5$, $n_1 = 7$, $n_2 = 7$, $n_3 = 7$, and $n_4 = 7$.
Step 4. We use the χ^2 distribution.
Step 5. Reject H_0 in favor of H_1 if the computed $\boldsymbol{H}$ value is > 11.345. Otherwise, fail to reject H_0.
Step 6. $\boldsymbol{H} = 6.13$.
Step 7. Since $6.13 < 11.345$, we fail to reject H_0. The mean number of errors made by those from different schools is the same.

8. Step 1. H_0: The mean sales for all four price groups are equal. H_1: At least one of the means is different.
Step 2. $\alpha = 0.05$.
Step 3. $N = 16$, $k = 4$, $R_1 = 23.5$, $R_2 = 35.5$, $R_3 = 41.5$, $R_4 = 35.5$, $n_1 = 4$, $n_2 = 4$, $n_3 = 4$, and $n_4 = 4$.
Step 4. We use the χ^2 distribution.
Step 5. Reject H_0 in favor of H_1 if the computed $\boldsymbol{H}$ value is > 7.815. Otherwise, fail to reject H_0.
Step 6. $\boldsymbol{H} = 1.92$.
Step 7. Since $1.92 < 7.815$, we fail to reject H_0. The mean number of sales at the different prices is equal.

10. Step 1. H_0: The mean study times are equal for all semesters. H_1: At least one of the means is different.
Step 2. $\alpha = 0.05$.
Step 3. $N = 24$, $k = 3$, $R_1 = 60$, $R_2 = 157.5$, $R_3 = 82.5$, $n_1 = 7$, $n_2 = 9$, and $n_3 = 8$.
Step 4. We use the χ^2 distribution.
Step 5. Reject H_0 in favor of H_1 if the computed $\boldsymbol{H}$ value is > 5.991. Otherwise, fail to reject H_0.
Step 6. $\boldsymbol{H} = 7.43$.
Step 7. Since $7.43 > 5.991$, we reject H_0. The mean is different for at least one of the semesters.

12. Step 1. H_0: The means for all groups are equal. H_1: At least one of the means is different.
Step 2. $\alpha = 0.05$.
Step 3. $N = 17$, $k = 3$, $R_1 = 74.5$, $R_2 = 24.5$, $R_3 = 54$, $n_1 = 8$, $n_2 = 4$, and $n_3 = 5$.
Step 4. We use the χ^2 distribution.
Step 5. Reject H_0 in favor of H_1 if the computed $\boldsymbol{H}$ value is > 5.991. Otherwise, fail to reject H_0.
Step 6. $\boldsymbol{H} = 1.96$.
Step 7. Since $1.96 < 5.991$, we fail to reject H_0. The mean time until relief is obtained is the same for the three brands of aspirin.

14. Step 1. H_0: The means for all groups are equal. H_1: At least one of the means is different.
Step 2. $\alpha = 0.01$.
Step 3. $N = 45$, $k = 5$, $R_1 = 211$, $R_2 = 278.5$, $R_3 = 136$, $R_4 = 128.5$, $R_5 = 281$, $n_1 = 10$, $n_2 = 11$, $n_3 = 8$, $n_4 = 5$, and $n_5 = 11$.
Step 4. We use the χ^2 distribution.
Step 5. Reject H_0 in favor of H_1 if the computed $\boldsymbol{H}$ value is > 14.86. Otherwise, fail to reject H_0.
Step 6. $\boldsymbol{H} = 2.85$.
Step 7. Since $2.85 < 14.86$, we fail to reject H_0. The mean time until relief is obtained is the same for the three brands of aspirin.

13-6 Runs Test for Randomness

2. The H_0 used in a runs test is that there is randomness in the sequential data. The H_1 is that there is an underlying pattern in the data.

4. The lower and upper critical values are 11 and 23. Since 10 (the number of runs) does not fall between these values, reject H_0. There is an underlying pattern in the data.

6. Step 1. H_0: There is randomness in the data. H_1: The data are not random; rather, there is a pattern in the data.
   Step 2. $\alpha = 0.05$.
   Step 3. There are 7 runs.
   Step 4. $n_1 = 10$ and $n_2 = 8$.
   Steps 5 and 6. Reject H_0 in favor of H_1 if the sample r value doesn't fall between 5 and 15.
   Step 7. Since 7 falls between 5 and 15, we fail to reject H_0. The data are random.

8. Step 1. H_0: There is randomness in answers to the test. H_1: The data are not random; rather, there is a pattern in the data.
   Step 2. $\alpha = 0.05$.
   Step 3. There are 20 runs.
   Step 4. $n_1 = 10$ and $n_2 = 10$.
   Steps 5 and 6. Reject H_0 in favor of H_1 if the sample r value doesn't fall between 6 and 16.
   Step 7. Since 20 does not fall between 6 and 16, we reject the H_0. The data are not random.

10. Step 1. H_0: The results of the coin tosses are random. H_1: The data are not random; rather, there is a pattern in the outcomes.
    Step 2. $\alpha = 0.05$.
    Step 3. There are 6 runs.
    Step 4. $n_1 = 7$ and $n_2 = 13$.
    Steps 5 and 6. Reject H_0 in favor of H_1 if the sample r value doesn't fall between 5 and 15.
    Step 7. Since 6 falls between 5 and 15, we fail to reject H_0. The data are random, and Gary is just unlucky.

12. Step 1. H_0: There is a randomness in judge's decisions. H_1: The decisions are not random; rather, there is a pattern.
    Step 2. $\alpha = 0.05$.
    Step 3. There are 9 runs.
    Step 4. $n_1 = 20$ and $n_2 = 6$.
    Steps 5 and 6. Reject H_0 in favor of H_1 if the sample r value doesn't fall between 6 and 14.
    Step 7. Since 9 falls between 6 and 14, we fail to reject H_0. The judge's decisions are in a random order.

13-7 Spearman Rank Correlation Coefficient

2. **(a)** The correlation coefficient is too close to 0 to be significant.
   (b) There's a strong negative correlation between the two rankings. High ranks in one category are paired with low ranks in the other.
   (c) There's an error here. The Spearman rank correlation can only range between -1.00 and $+1.00$.

4. $t = 0.58\sqrt{\dfrac{12-2}{1-0.58^2}} = 2.2515$. The critical t values for the distribution with 10 df and $\alpha = 0.01$ are $+3.169$. Since the $t = 2.2515$ falls between ± 3.169, we fail to reject H_0 and conclude that there is no significant correlation.

6. With $\Sigma D^2 = 5366$, and $n = 14$, $r_s = -0.178$.
Step 1. H_0: $p_s = 0$ and H_1: $p_s \neq 0$.
Step 2. $\alpha = 0.01$.
Step 3. We use a t distribution.
Step 4. With 12 df and $\alpha = 0.01$, the critical t values are ± 3.055.
Step 5. Reject H_0 in favor of H_1 if $t < -3.055$ or if $t > 3.055$. Otherwise, fail to reject H_0.
Step 6. $t = -0.6267$.
Step 7. Since -0.6267 falls in the acceptance region, we fail to reject H_0. There is no significant correlation in the rankings of the teams in the two time periods.

8. With $\Sigma D^2 = 1413.5$, and $n = 21$, $r_s = 0.0821$.
Step 1. H_0: $p_s = 0$ and H_1: $p_s > 0$.
Step 2. $\alpha = 0.05$.
Step 3. We use a t distribution.
Step 4. The critical t value is 1.729.
Step 5. Reject H_0 in favor of H_1 if $t > 1.729$. Otherwise, fail to reject H_0.
Step 6. $t = 0.3593$.
Step 7. Since $0.3593 < 1.729$, we fail to reject H_0. There is no significant correlation.

10. With $\Sigma D^2 = 9$, and $n = 8$, $r_s = 0.8929$.

12. With $\Sigma D^2 = 78$, and $n = 10$, $r_s = 0.5273$.
Step 1. H_0: $p_s = 0$ and H_1: $p_s > 0$.
Step 2. $\alpha = 0.05$.
Step 3. We use a t distribution.
Step 4. The critical t value is 1.860
Step 5. Reject H_0 in favor of H_1 if $t > 1.860$. Otherwise, fail to reject H_0.
Step 6. $t = 1.76$.
Step 7. Since $1.76 < 1.860$, we reject H_0. There's no significant correlation between the cost of a film and its gross receipts.

14. With $\Sigma D^2 = 51$, and $n = 8$, $r_s = 0.3929$.

16. With $\Sigma D^2 = 110$, and $n = 25$, $r_s = 0.958$.
Step 1. H_0: $p_s = 0$ and H_1: $p_s \neq 0$.
Step 2. $\alpha = 0.01$.
Step 3. We use a t distribution.
Step 4. The critical t values are ± 2.807.
Step 5. Reject H_0 in favor of H_1 if $t < -2.807$ or if $t > 2.807$. Otherwise, fail to reject H_0.
Step 6. $t = 15.96$.
Step 7. Since $15.96 > 2.807$, we reject H_0. There is a significant positive correlation.

Answers to Even-Numbered Exercises

2. Step 1. H_0: Before and after Eficholp shows no difference. H_1: Eficholp is effective.

Step 2. $\alpha = 0.05$.

Step 3. Compute the difference and count the number of differences $\neq 0$.

Person	Level Before Use	Level After Use	Difference
A	263	214	49
B	194	188	6
C	273	284	−11
D	185	185	0
E	238	264	−26
F	212	190	22
G	189	185	4
H	164	153	11
I	248	248	0
J	261	229	32

The number of difference is $n = 8$.

Step 4. Use the one-tailed T statistic test. With $n = 8$, the table value of T is 5.

Step 5. Reject H_0 in favor of H_1 if the computed $\boldsymbol{T}$ value $\leq$ the table T value. Otherwise fail to reject H_0.

Step 6. Complete the table and compute $\boldsymbol{T}$.

Person	Level Before Use	Level After Use	Difference	Rank	Sign Rank Positive	Sign Rank Negative
A	263	214	49	8.0	8.0	
B	194	188	6	2.0	2.0	
C	273	284	−11	3.5		−3.5
D	185	185	0	ignore		
E	238	264	−26	6.0		−6.0
F	212	190	22	5.0	5.0	
G	189	185	4	1.0	1.0	
H	164	153	11	3.5	3.5	
I	248	248	0	ignore		
J	261	229	32	7.0	7.0	
					26.5	−9.5

For this right-tailed test, $\boldsymbol{T}$ is the sum of the negative ranks, thus $\boldsymbol{T} = 9.5$.

Step 7. Since the computed $\boldsymbol{T}$ value 9.5 is greater than the table value of 5, we fail to reject H_0.

4. Step 1. H_0: The mean weights of the 4 groups are equal. H_1: The mean weights of the 4 groups are not equal.

Step 2. $\alpha = 0.05$.

Step 3. Rank the data irrespective of sample category.

Group 1	Rank	Group 2	Rank	Group 3	Rank	Group 4	Rank
84	17	56	4.5	70	11	47	2
93	18	78	14.5	59	6	73	13
83	16	56	4.5	78	14.5	104	20.5
61	7.5	61	7.5	53	3	71	12
121	23			104	20.5	69	10
67	9			110	22	99	19
				40	1		
$n = 6$	90.5	$n = 4$	31.0	$n = 7$	78.0	$n = 6$	76.5

Step 4. Use the one-tailed χ^2 test, with $n = 3$ the H critical value is 7.81.

Step 5. Reject H_0 in favor of H_1 if the computed $\boldsymbol{H}$ value > 7.81. Otherwise fail to reject H_0.

Step 6. Compute the H value using the formula:

$$\boldsymbol{H} = \frac{12}{N(N+1)}\left(\frac{R_1^2}{n_1} + \frac{R_2^2}{n_2} + \frac{R_3^2}{n_3} + \frac{R_4^2}{n_4}\right) - 3(N+1)$$

$$= \frac{12}{23(23+1)}\left(\frac{90.5^2}{6} + \frac{31^2}{4} + \frac{78^2}{7} + \frac{76.5^2}{6}\right) - 3(23+1)$$

$$= 0.0217(3449.81) - 72 = 74.99 - 72 = 2.99.$$

Step 7. Since the computed H value 2.99 is not greater than the value of 7.81, we fail to reject H_0.

This is confirmed by MINITAB.

```
MTB > Kruskal-Wallis c1 c2.

Kruskal-Wallis Test

Kruskal-Wallis Test on Weight

Group        N     Median     Ave Rank          Z
1            6      83.50         15.1       1.30
2            4      58.50          7.8      -1.38
3            7      70.00         11.1      -0.40
4            6      72.00         12.8       0.32
Overall     23                    12.0

H = 3.00  DF = 3  P = 0.392
H = 3.00  DF = 3  P = 0.391 (adjusted for ties)
```

6. Step 1. H_0: The number of errors made by a random sample of individuals in each group are equal, and H_1: The number of errors made are not equal.

Step 2. $\alpha = 0.02$.

Step 3. Rank the data irrespective of sample category.

Chronic Schizophrenics	Rank	Normal Comparison Subjects	Rank
8	5	5	1
6	2	7	3
11	11	9	7.5
12	12	10	9.5
8	5	8	5
10	9.5		
9	7.5		
$n_1 = 7$	$R_1 = 52.0$	$n_2 = 5$	$R_2 = 26.0$

Step 4. Use the two-tailed $\boldsymbol{U}$ statistic test. With $n_1 = 7$, $n_2 = 5$, the table value of U is 3.

Step 5. Reject H_0 in favor of H_1 if the computed $\boldsymbol{U}$ value $\leq$ the table U value. Otherwise fail to reject H_0.

Step 6. Compute U_1 and U_2.

$$U_1 = n_1(n_2) + \frac{n_1(n_1+1)}{2} - R_1 = 7 \cdot 5 + \frac{7 \cdot 8}{2} - 52 = 11.$$

$$U_2 = n_1(n_2) + \frac{n_2(n_2+1)}{2} - R_2 = 7 \cdot 5 + \frac{5 \cdot 6}{2} - 26 = 24.$$

The $\boldsymbol{U}$ statistic is 11, the lesser of 11 and 24.

Step 7. Since the computed $\boldsymbol{U}$ value 11 is greater than the table value of 3, we fail to reject H_0.

8. Step 1. H_0: There is no difference between before and after bonus plan. H_1: Workers show improvement after bonus plan.

Step 2. $\alpha = 0.05$.

Step 3. Compute the difference and count the number of differences $\neq 0$.

Worker	Output Before	Output After	Difference
Harris Tweed	80	85	5
Stitch N. Thyme	75	75	0
Les Brown	65	71	6
Mary Taylor	82	79	−3
Chuck Moore	56	68	12
Tex Tyle	70	86	16
Ray Ohn	73	71	−2
Terry Clothe	62	59	−3

The number of difference is $n = 7$.

Step 4. Use the one-tailed T statistic test. With $n = 7$, the table value of T is 3.

Step 5. Reject H_0 in favor of H_1 if the computed $\boldsymbol{T}$ value $\leq$ the table T value. Otherwise, fail to reject H_0.

Step 6. Complete the table and compute $\boldsymbol{T}$.

Worker	Output Before	Output After	Difference	Rank	Sign Positive	Rank Negative
Harris Tweed	80	85	5	4.0	4.0	
Stitch N. Thyme	75	75	0	ignore		
Les Brown	65	71	6	5.0	5.0	
Mary Taylor	82	79	−3	2.5		−2.5
Chuck Moore	56	68	12	6.0	6.0	
Tex Tyle	70	86	16	7.0	7.0	
Ray Ohn	73	71	−2	1.0		−1.0
Terry Clothe	62	59	−3	2.5		−2.5
					22	−6

For this right-tailed test, $\boldsymbol{T}$ is the sum of the negative ranks, thus $\boldsymbol{T} = 6$.

Step 7. Since the computed $\boldsymbol{T}$ value 6 is greater than the table value of 3, we fail to reject H_0.

10. Step 1. H_0: The mean hearing level is the same for all age groups. H_1: The mean hearing levels are not equal.

Step 2. $\alpha = 0.05$.

Step 3. Rank the data irrespective of sample category.

Thirties	Rank	Fifties	Rank	Sixties	Rank	Seventies	Rank
9	7.5	9	7.5	19	14	18	13
13	10	5	3	8	5.5	22	15
5	3	8	5.5	14	11	24	16
5	3	3	1	26	17		
10	9	15	12				
$n = 5$	32.5	$n = 5$	29.0	$n = 4$	47.5	$n = 3$	44

Step 4. Use the one-tailed χ^2 test, with $n = 3$ the H statistic 7.81.

Step 5. Reject H_0 in favor of H_1 if the computed $\boldsymbol{H}$ value > 7.81. Otherwise fail to reject H_0.

Step 6. Compute the $\boldsymbol{H}$ value using the formula:

$$H = \frac{12}{N(N+1)}\left(\frac{R_1^2}{n_1} + \frac{R_2^2}{n_2} + \frac{R_3^2}{n_3} + \frac{R_4^2}{n_4}\right) - 3(N+1)$$

$$= \frac{12}{17(17+1)}\left(\frac{32.5^2}{5} + \frac{29^2}{5} + \frac{47.5^2}{4} + \frac{44^2}{3}\right) - 3(17+1)$$

$$= 0.0392(1,588.846) - 54 = 52.28 - 54 = 8.28.$$

Step 7. Since the computed $\boldsymbol{H}$ value 8.28 is greater than the value of 7.81, we reject H_0.

Using MINITAB we get:

```
MTB > Kruskal-Wallis 'Tone' 'Age'.
```

```
Kruskal-Wallis Test

Kruskal-Wallis Test on Tone

Age          N     Median     Ave Rank         Z
30           5      9.000          6.5     -1.32
50           5      8.000          5.8     -1.69
60           4     16.500         11.9      1.30
70           3     22.000         14.7      2.14
Overall     17                     9.0

H = 8.31   DF = 3   P = 0.040
H = 8.37   DF = 3   P = 0.039 (adjusted for ties)
```

12. Step 1. H_0: There is no difference in the tests scores. H_1: The second test scores were higher than those in the first test scores.
Step 2. $\alpha = 0.05$.
Step 3. Compute the difference and count the number of differences $\neq 0$.

Test 1	Test 2	Difference
12	19	7
12	11	−1
10	10	0
16	16	0
14	16	2
14	14	0
12	11	−1
17	14	−3
16	18	2
13	14	1
16	14	−2

The number of difference is $n = 8$.
Step 4. Use the one-tailed T statistic test. With $n = 8$, the table value of T is 5.
Step 5. Reject H_0 in favor of H_1 if the computed $\boldsymbol{T}$ value $\leq$ the table T value. Otherwise fail to reject H_0.
Step 6. Complete the table and compute $\boldsymbol{T}$.

Test 1	Test 2	Difference	Rank	Sign Positive	Rank Negative
12	19	7	8	8	
12	11	−1	2		−2
10	10	0	ignore		
16	16	0	ignore		
14	16	2	5	5	
14	14	0	ignore		
12	11	−1	2		−2
17	14	−3	7		−7
16	18	2	5	5	
13	14	1	2	2	
16	14	−2	5		−5
				20	−16

For this right-tailed test, $\boldsymbol{T}$ is the sum of the negative ranks, thus $\boldsymbol{T} = 16$.
Step 7. Since the computed $\boldsymbol{T}$ value 16 is greater than the table value of 5, we fail to reject H_0.

14. Step 1. H_0: $p = 0.5$ and H_1: $p > 0.5$.
Step 2. $\alpha = 0.05$.

Step 3. 43 improved their gpa, 21 saw their gpa decrease, and the rest, 11, found that their gpa remained the same.

Step 4. Since $n = 64$, we use the z distribution. This is a one-tailed test, with $\alpha = 0.05$ the critical z value is 1.645.

Step 5. Reject H_0 in favor of H_1 if $z > 1.645$.

Step 6. $z = \dfrac{2r - n}{\sqrt{n}} = \dfrac{2(43) - 64}{\sqrt{64}} = 2.75.$

Step 7. Since $2.75 > 1.645$, we reject H_0. The newprogram increases an athlete's gpa.

16. Step 1. H_0: $p = 0.5$ and H_1: $p \neq 0.5$.

Step 2. $\alpha = 0.05$.

Step 3. Tally the sign difference.

Device 1	Device 2	Sign
10.51	11.97	+
20.30	19.47	−
27.88	30.19	+
40.56	38.43	−
47.57	46.23	−
58.35	57.15	−
66.63	66.27	−

5 minus, 2 pluses.

Step 4. Use the binomial probability table with $n = 7$, $r = 2$, and $p = 0.5$.

Step 5. Reject H_0 in favor of H_1 if $0.05 >$ the probability of the sample results.

Step 6. $P(0) + P(1) + P(2) = 0.0078 + 0.0547 + 0.1641 = 0.2266.$

Step 7. Since $0.2266 > 0.05$, we fail to reject H_0. The measurements from the devices are the same.

18. Step 1. H_0: $p_s = 0$ and H_1: $p_s \neq 0$.

Step 2. $\alpha = 0.05$.

Step 3. Use the t distribution.

Step 4. With $15 - 2 = 13$ df the critical t value is about 2.160.

Step 5. Reject H_0 in favor of H_1 if $t > 2.160$. Otherwise, fail to reject H_0.

Step 6. Compute r_s and t.

Day	Temperature Rank	Sales Rank	Difference between Ranks, D	D^2
1	6	5	1	1
2	11	12	−1	1
3	4	2	2	4
4	7	7	0	0
5	1	4	−3	9
6	12	14	−2	4
7	8	10	−2	4
8	2	1	1	1
9	15	15	0	0
10	14	13	1	1
11	5	3	2	4
12	10	9	1	1
13	13	11	2	4
14	9	8	1	1
15	3	6	−3	9
			$\Sigma D = 0$	$\Sigma D^2 = 44$

$$r_s = 1 - \frac{6\,\Sigma D^2}{n(n^2-1)} = 1 - \frac{6(44)}{15(225-1)} = 1 - 0.0786 = 0.9214.$$

$$t = r_s \cdot \sqrt{\frac{n-2}{1-r_s^2}} = 0.9214 \cdot \sqrt{\frac{13}{1-0.8490}} = 8.55.$$

Step 7. Since 8.55 is greater than the critical t value 2.160, we reject H_0. There is a strong relationship between daily temperature and sales.

20. Step 1. H_0: There is no difference in career decision-making skills among various ethnic groups. H_1: There is a difference in career decision-making skills among various ethnic groups.

Step 2. $\alpha = 0.01$.

Step 3. Rank the data irrespective of sample category.

African-American	Rank	Hispanic	Rank	Caucasian	Rank
17	15	12	3.5	13	6
9	1	10	2	14	9
13	6	15	12	14	9
16	14	13	6	15	12
12	3.5			15	12
				14	9
$n = 5$	39.5	$n = 4$	23.5	$n = 6$	57

Step 4. Use the one-tailed χ^2 test with 2 df the H statistic 9.21.

Step 5. Reject H_0 in favor of H_1 if the computed $\boldsymbol{H}$ value > 9.21. Otherwise fail to reject H_0.

Step 6. Compute the $\boldsymbol{H}$ value using the formula:

$$\boldsymbol{H} = \frac{12}{N(N+1)}\left(\frac{R_1^2}{n_1} + \frac{R_2^2}{n_2} + \frac{R_3^2}{n_3}\right) - 3(N+1)$$

$$= \frac{12}{15(15+1)}\left(\frac{39.5^2}{5} + \frac{23.5^2}{4} + \frac{57^2}{6}\right) - 3(15+1)$$

$$= 0.05(991.6125) - 48 = 49.58 - 48 = 1.58.$$

Step 7. Since the computed $\boldsymbol{H}$ value 1.58 is less than the value of 9.21, we fail to reject H_0.

22. Step 1. H_0: $p^s = 0$ and H_1: $p_s \neq 0$.

Step 2. $\alpha = 0.05$.

Step 3. Use the t distribution.

Step 4. With $20 - 2 = 18$ df and a two-tailed test, the critical t value is about ± 2.101.

Step 5. Reject H_0 in favor of H_1 if $\boldsymbol{t} > 2.101$ or if $\boldsymbol{t} < -2.101$. Otherwise, fail to reject H_0.

Step 6. Compute r_s and $\boldsymbol{t}$.

Research Rank	Primary Care Rank	Difference between Ranks, D	D^2
1	17	−16	256
2	2	0	0
3	18	−15	225
4	4	0	0
5	19	−14	196
6	20	−14	196
7	1	6	36
8	8	0	0
9	7	2	4
10	14	−4	16
11	3	8	64
12	15	−3	9
13	9	4	16
14	13	1	1
15	6	9	81
16	10	6	36
17	11	6	36
18	12	6	36
19	5	14	196
20	16	4	16
		$\Sigma D = 0$	$\Sigma D^2 = 1420$

$$r_s = 1 - \frac{6\,\Sigma D^2}{n(n^2-1)} = 1 - \frac{6(1420)}{20(400-1)} = 1 - 1.068 = -0.068.$$

$$t = r_s \cdot \sqrt{\frac{n-2}{1-r_s^2}} = -0.068 \cdot \sqrt{\frac{18}{1-0.004624}} = -0.289.$$

Step 7. Since −0.289 is between the critical t values ±2.101, we fail to reject H_0.

Using MINITAB we get:

```
MTB > Correlation  'Research' 'Primary Care'.

Correlations (Pearson)

Correlation of Research and Primary Care = -0.068, P-Value = 0.777
```

24. Step 1. H_0: There is no weight difference using Slender Fast. H_1: There is a significant difference in weight due to Slender Fast.

Step 2. $\alpha = 0.05$.

Step 3. Compute the difference and count the number of differences $\neq 0$.

Subject	Before	After	Difference
1	135	115	−20
2	167	165	−2
3	205	163	−42
4	115	121	6
5	175	148	−27
6	134	141	7
7	110	110	0

The number of difference is $n = 6$.

Step 4. Use the one-tailed T statistic test. With $n = 6$, the table value of T is 2.

Step 5. Reject H_0 in favor of H_1 if the computed $\boldsymbol{T}$ value $\leq$ the table T value. Otherwise fail to reject H_0.

Step 6. Complete the table and compute T.

Subject	Before	After	Difference	Rank	Sign Positive	Rank Negative
1	135	115	−20	4		−4
2	167	165	−2	1		−1
3	205	163	−42	6		−6
4	115	121	6	2	2	
5	175	148	−27	5		−5
6	134	141	7	3	3	
7	110	110	0	ignore		
					5	−16

For this left-tailed test, T is the sum of the positive ranks, thus $T = 5$.

Step 7. Since the computed T value 5 is greater than the table value of 2, we fail to reject H_0.

26. Step 1. H_0: The test score are equal for all four versions. H_1: The test score are not equal for all four versions.

Step 2. $\alpha = 0.05$.

Step 3. Rank the data irrespective of sample category.

Version 1	Rank	Version 2	Rank	Version 3	Rank	Version 4	Rank
59	4.5	88	19.5	70	13.0	77	16.0
93	22.5	62	7.0	59	4.5	73	15.0
87	18.0	56	3.0	68	11.0	94	24.0
66	9.0	61	6.0	53	2.0	71	14.0
91	21.0	88	19.5	64	8.0	69	12.0
67	10.0	93	22.5	80	17.0	99	25.5
99	25.5			40	1.0		
$n = 7$	110.5	$n = 6$	77.5	$n = 7$	56.5	$n = 6$	106.5

Step 4. Use the one-tailed χ^2 test with 3 df the H statistic 7.81.

Step 5. Reject H_0 in favor of H_1 if the computed H value > 7.81. Otherwise fail to reject H_0.

Step 6. Compute the H value using the formula:

$$\boldsymbol{H} = \frac{12}{N(N+1)}\left(\frac{R_1^2}{n_1} + \frac{R_2^2}{n_2} + \frac{R_3^2}{n_3} + \frac{R_4^2}{n_4}\right) - 3(N+1)$$

$$= \frac{12}{26(26+1)}\left(\frac{110.5^2}{7} + \frac{77.5^2}{6} + \frac{56.5^2}{7} + \frac{106.5^2}{6}\right) - 3(26+1)$$

$$= 0.0171(6836.095) - 81 = 116.90 - 81 = 35.9.$$

Step 7. Since the computed H value 10.35 is greater than the table value 7.81, we reject H_0.

28. Step 1. H_0: $p = 0.5$ and H_1: $p > 0.5$.

Step 2. $\alpha = 0.05$.

Step 3. 25 of these stocks had increased in value, 34 had decreased in value, and the rest had stayed the same.

Step 4. Since $n = 59$ we use the z distribution. This is a one-tailed test, with $\alpha = 0.05$ the critical z value is 1.645.

Step 5. Reject H_0 in favor of H_1 if $z > 1.645$.

Step 6. $z = \dfrac{2r - n}{\sqrt{n}} = \dfrac{2(34) - 59}{\sqrt{59}} = 1.1717.$

Step 7. Since $1.1717 < 1.645$, we fail to reject H_0.

30. Step 1. H_0: There is randomness in the data sequence, and H_1: The data are not random.

Step 2. $\alpha = 0.05$.

Step 3. Count the number of runs.

g	b	g	g	b	b	b	g	b	g	g	b	b	g	b
1	2		3			4	5	6		7		8	9	10

Hence r (the number of runs) is 10.

Step 4. Calculate the number of girls, $n_1 = 7$, and the number of boys, $n_2 = 8$.

Step 5. Reject H_0 in favor of H_1 if the sample r value is not between a and b. Otherwise, fail to reject H_0.

Step 6. Using Appendix 9 with $n_1 = 7$ and $n_2 = 8$ we get the values $a = 4$ and $b = 13$.

Step 7. Since $a < 10 < b$, we fail to reject H_0.

32. Step 1. H_0: The TV viewing habits are the same for students in different middle school grades. H_1: The TV viewing habits are not the same.

Step 2. $\alpha = 0.05$.

Step 3. Rank the data irrespective of sample category.

Sixth Grade	Rank	Seventh Grade	Rank	Eighth Grade	Rank
459	12	115	3	272	8
311	10	153	5	88	2
152	4	201	7	374	11
293	9	30	1	178	6
$n = 4$	35	$n = 4$	16	$n = 4$	27

Step 4. Use the one-tailed χ^2 test with 2 df the H statistic is 5.99.

Step 5. Reject H_0 in favor of H_1 if the computed $\boldsymbol{H}$ value > 5.99. Otherwise fail to reject H_0.

Step 6. Compute the $\boldsymbol{H}$ value using the formula:

$$\boldsymbol{H} = \frac{12}{N(N+1)}\left(\frac{R_1^2}{n_1} + \frac{R_2^2}{n_2} + \frac{R_3^2}{n_3}\right) - 3(N+1)$$

$$= \frac{12}{12(12+1)}\left(\frac{35^2}{4} + \frac{16^2}{4} + \frac{27^2}{4}\right) - 3(12+1)$$

$$= 0.0769(552.5) - 39 = 42.49 - 39 = 3.49.$$

Step 7. Since the computed $\boldsymbol{H}$ value 3.49 is less than the table value 5.99, we fail to reject H_0.

34. Step 1. H_0: There is randomness in the data sequence, and H_1: The data are not random.

Step 2. $\alpha = 0.05$.

Step 3. Count the number of runs.

T	C	T	T	T	C	C	T	C	T	T	T	T	T	C
1	2			3		4	5	6					7	8

Hence r (the number of runs) is 8.

Step 4. Calculate the number of T's, $n_1 = 10$, and the number of C's, $n_2 = 5$.

Step 5. Reject H_0 in favor of H_1 if the sample r value is not between a and b. Otherwise, fail to reject H_0.

Step 6. Using Appendix 9 with $n_1 = 10$ and $n_2 = 5$ we get the values $a = 3$ and $b = 12$.

Step 7. Since $a < 8 < b$, we fail to reject H_0.

36. Step 1. H_0: The means for the two groups are equal, and H_1: The means are not equal.

Step 2. $\alpha = 0.05$.

Step 3. Rank the data irrespective of sample category.

Chronic Schizophrenics	Rank	Normal Subjects	Rank
87	3	118	9
91	4	107	5
116	7.5	110	6
86	2	83	1
116	7.5		
$n_1 = 5$	$R_1 = 24$	$n_2 = 4$	$R_2 = 21$

Step 4. Use the two-tailed U statistic test. With $n_1 = 5$, $n_2 = 4$, the table value of U is 2.

Step 5. Reject H_0 in favor of H_1 if the computed $\boldsymbol{U}$ value $\leq$ the table U value. Otherwise fail to reject H_0.

Step 6. Compute U_1 and U_2.

$$U_1 = n_1 n_2 + \frac{n_1(n_1+1)}{2} - R_1 = 5 \cdot 4 + \frac{5 \cdot 6}{2} - 24 = 11$$

$$U_2 = n_1 n_2 + \frac{n_2(n_2+1)}{2} - R_2 = 5 \cdot 4 + \frac{4 \cdot 5}{2} - 21 = 9$$

The $\boldsymbol{U}$ statistic is 9, the lesser of 11 and 9.

Step 7. Since the computed $\boldsymbol{U}$ value 9 is greater than the table value of 2, we fail to reject H_0.

38. Step 1. H_0: The means for the two groups are equal, and H_1: The High School A is more aggressive.

Step 2. $\alpha = 0.05$.

Step 3. Rank the data irrespective of sample category.

School A	Rank	School B	Rank
43	10	47	11
56	13	68	14
31	3	39	7
30	2	29	1
41	8	36	5
38	6	42	9
		33	4
		54	12
$n_1 = 6$	$R_1 = 42$	$n_2 = 8$	$R_2 = 63$

Step 4. Use the one-tailed U statistic test. With $n_1 = 6$, $n_2 = 8$, the table value of U is 10.

Step 5. Reject H_0 in favor of H_1 if the computed U value $\leq$ the table U value. Otherwise fail to reject H_0.

Step 6. Compute U_1. $U_1 = n_1 n_2 + \dfrac{n_1(n_1+1)}{2} - R_1 = 6 \cdot 8 + \dfrac{6 \cdot 7}{2} - 42 = 27$.

Step 7. Since the computed $\boldsymbol{U}$ value 27 is greater than the table value of 10, we fail to reject H_0.

40. Step 1. H_0: The mean weights of the 4 groups are equal. H_1: The mean weights of the 4 groups are not equal.

Step 2. $\alpha = 0.05$.

Step 3. Rank the data irrespective of sample category.

Group 1	Rank	Group 2	Rank	Group 3	Rank	Group 4	Rank
27	2.5	39	12	37	10	24	1
50	18	36	7.5	28	4	53	22
43	13	47	17	44	14.5	51	20
31	6	51	20	36	7.5	51	20
37	10			30	5	45	16
37	10			27	2.5	65	23
				44	14.5		
$n = 6$	59.5	$n = 4$	56.5	$n = 7$	58.0	$n = 6$	102.0

Step 4. Using the one-tailed χ^2 test with 3 df we find the H statistic 7.81.

Step 5. Reject H_0 in favor of H_1 if the computed $\boldsymbol{H}$ value > 7.81. Otherwise fail to reject H_0.

Step 6. Compute the $\boldsymbol{H}$ value using the formula:

$$\boldsymbol{H} = \frac{12}{N(N+1)}\left(\frac{R_1^2}{n_1}+\frac{R_2^2}{n_2}+\frac{R_3^2}{n_3}+\frac{R_4^2}{n_4}\right) - 3(N+1)$$

$$= \frac{12}{23(23+1)}\left(\frac{59.2^2}{6}+\frac{56.5^2}{4}+\frac{58^2}{7}+\frac{102^2}{6}\right) - 3(23+1)$$

$$= 0.0217(3602.676) - 72 = 6.178.$$

Step 7. Since the computed $\boldsymbol{H}$ value 6.178 is less than the value of 7.81, we fail to reject H_0.

42. Step 1. H_0: There is randomness in the data sequence, and H_1: The data are not random.

Step 2. $\alpha = 0.05$.

Step 3. Count the number of runs.

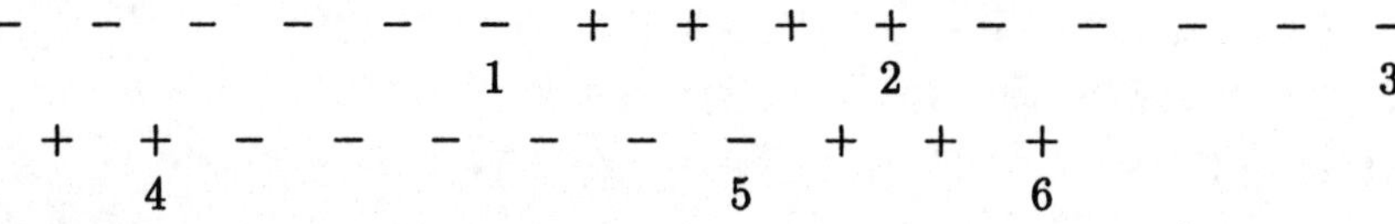

Hence r (the number of runs) is 6.

Step 4. Calculate the number of −'s, $n_1 = 20$, and the number of +'s, $n_2 = 9$.

Step 5. Reject H_0 in favor of H_1 if the sample r value is not between a and b. Otherwise, fail to reject H_0.

Step 6. Using Appendix 9 with $n_1 = 20$ and $n_2 = 9$ we get the values $a = 8$ and $b = 18$.

Step 7. Since $6 < a$, we reject H_0.

Unit 2

T 1. An average is a descriptive statistic.

T 2. Whenever any large mass of data is condensed so that the summary results are used for reporting purposes, descriptive statistics have been computed.

F 3. Since statistical inference involves drawing conclusions about a population based on sample information, no descriptive statistics are used.

F 4. A census simply refers to a complete analysis of the entire sample.

F 5. Since sampling techniques are never perfect, it is impossible to test assumptions about an unknown population characteristic.

T 6. In statistical inference, the sample statistics are determined objectively, but the interpretation of these statistics may vary from person to person.

T 7. One of the objectives in statistics is to provide measures which identify and describe relationships between variables.

F 8. The difference between inferential and descriptive statistics is that inferential statistics does not rely on summary measures of the sample to draw conclusions.

F 9. An average is a measure of dispersion.

T 10. A judgment sample is a sample based on someone's expertise about the population.

T 11. A census is nothing but a sample of 100 percent of the population.

F 12. Primary data is preferred over secondary data because it contains no errors.

F 13. Coding of data reduces the ability to make decisions because the original data has been abbreviated.

F 14. A parameter, since it is a population characteristic, is the maximum value that a statistic may assume.

T 15. Coding is a method used to classify data.

T 16. Measures of dispersion refer to the spread of the data about the central measure.

F 17. Samples always completely represent the population.

T 18. The word statistics, in a singular sense, refers to a subject of study.

F 19. The subset of the total group that supplies incomplete data to a decision maker is called a census.

F 20. A population characteristic or measure is called a statistic.

F 21. A sample characteristic or measure is called a parameter.

T 22. External data are supplied by publications such as Science and Business Week.

T 23. The statistical inference process involves the use of a known sample statistic to arrive at a judgment about an unknown population parameter.

F 24. Data from secondary sources is always better to use because the descriptive statistics have usually already been done.

T 25. Common data-gathering practices make use of personal interviews and mail questionnaires.

T 26. Two software tools commonly used for data analysis are electronic spreadsheet and statistical analysis packages.

T 27. A spreadsheet is a program that accepts data values and relationships in the columns and rows of its worksheet.

F 28. Templates can be used with statistical analysis programs, but they aren't available for spreadsheets.

T 29. Statistical analysis packages are preprogrammed with the specialized formulas a user may need to carry out a range of statistical studies.

F 30. All statistical analysis programs were first written for large computer systems and then adapted for use with personal computers.

1. Which of the following does not characterize the procure in descriptive statistics?
   (a) Presenting quantitative facts.
   (b) Summarizing quantitative facts.
   (c) Collecting quantitative facts.
X (d) Interpreting quantitative facts.
   (e) Classifying quantitative facts.

2. In statistical inference procedures, the computed measures taken from samples are used:
   (a) to summarize decisions.
X (b) to provide a basis for decision making.
   (c) to eliminate uncertainly in decisions.
   (d) to classify decision makers.
   (e) to present the optimum decision.

3. Statistical inference involves drawing conclusions about:
   (a) the sample based on sample statistics.
   (b) the descriptive accuracy of quantitative facts.
   (c) the value of sample results based on population results.
X (d) the population characteristics based on sample information.
   (e) none of the above.

4. One of the reasons for selecting a sample is:
   (a) to provide summary measures of the population.
X (b) to provide estimates of population characteristics.
   (c) to provide information for riskless decisions about a population segment.
   (d) to eliminate all errors in judgments.
   (e) to provide information on the effect of sampling on selected segments.

5. A sample is sometimes desired over a census because a sample:
   (a) can provide error-free information.
   (b) has characteristics which can be used to determine the exact nature of population characteristics.
X (c) allows cost savings.
   (d) provides more accurate information than a census and also saves time and money.
   (e) will always result in less error than a census.

6. Statistical concepts and techniques are used to
   (a) organize data.
   (b) measure data.
   (c) evaluated data.
X (d) all of the above.
   (e) none of the above.

7. Which of the following is not a basic step in statistical problem solving?
   (a) Understanding and correctly defining the problem.
   (b) Gathering facts relevant to the problem.
   (c) Classifying and summarizing data.
   (d) Interpreting results.
X (e) Following up on decisions.

8. Primary sources are preferred over secondary sources because:
   (a) primary sources are free of error.
   (b) organizations which republish data are always unreliable.
   (c) primary sources are always cheaper than secondary sources.
X (d) secondary sources are subject to reproduction error and may not explain methodology or limitations of the data
   (e) None of the above are correct.

9. Which of the following is the first step in statistical problem solving methodology?
   (a) Arrangement for data collection services.
   (b) Use of census information to formulate a sampling plan for the project.
   (c) Identification of alternatives.
   (d) Development of the framework for data analysis.
X (e) None of the above.

10. Which of the following is not involved in the analysis of data?
   (a) Interpretation of descriptive statistics.
   (b) Making inferences from the statistics.
   (c) Use of the decision maker's experience.
   (d) Employment of the statistical aids.
X (e) All of the above are involved.

11. Statistics as a subject of study seeks to provide measures which
   (a) are used only for descriptive purposes.
   (b) shed light only on the sample characteristics.
   (c) prove undisputed cause and effect.
X (d) identify and describe relationships between variables.
   (e) are always superior to subjective judgments.

12. Which of the following best describes the relationship between statistics and decision making?
   (a) The decision maker should always follow the dictates of statistical results.
   (b) Statistics eliminates the need for subjective decisions.
   (c) Statistics eliminates all the uncertainty in decisions.
   (d) Decision making should not use descriptive statistics.
X (e) Statistics reduce uncertainty in decision making, but the proper use of statistics depends on the skill of the decision maker.

13. What should be the first step in statistical problem solving?
    (a) Analyzing data.
    (b) Identification of primary sources.
    (c) Creation of a problem.
X (d) Identification of the problem.
    (e) None of the above.

14. Which of the following is a benefit for the researcher who gathers new data for analysis?
    (a) The researcher may define the variables.
    (b) The researcher knows how the variables are measured.
    (c) The data are recent.
    (d) The methods for obtaining data are known.
X (e) All of the above are benefits.

15. The portion or subset of the population that supplies data to a decision maker is called a
X (a) sample.
    (b) parameter.
    (c) data item.
    (d) census.
    (e) none of the above.

16. A population characteristic or measure is called a
    (a) statistic.
    (b) census.
X (c) parameter.
    (d) universe.
    (e) none of the above.

17. Which of the following is not a probability sample?
    (a) Simple random sample.
    (b) Systematic sample.
    (c) Cluster sample.
X (d) Judgment sample.
    (e) None of the above.

18. "The statistical inference process involves the use of a known sample ______ to arrive at a judgment about an unknown population ______." The correct terms needed to complete the preceding sentence are:
X (a) statistic, parameter.
    (b) parameter, statistic.
    (c) census, statistic.
    (d) parameter, census.
    (e) hypothesis, statistic.

19. Unlike spreadsheets, statistical analysis packages
    (a) accept data from other sources.
X   (b) are preprogrammed with specialized formulas and built-in procedures a user may need to carry out a range of statistical studies.
    (c) add or remove data items, columns, or rows.
    (d) sort, merge, and manipulate facts in numerous ways.
    (e) convert numeric data into charts and graphs.

20. Which of the following isn't a statistical analysis package?
    (a) Mystat.
    (b) Minitab.
X   (c) SOS.
    (d) SPSS.
    (e) SAS.

1. Explain the process of statistical inference.

2. When should stratified sampling be performed?

3. A cable television company is considering adding a new shopping channel to their basic cable service. To help them decide, they call 500 customers and ask their opinion. Identify the population, the sample, parameter, and statistic for this situation.

4. A vintner needs to pick grapes when the sugar content reaches 23%. To tell if the grapes are really in their vineyard, she selects 10 bunches of grapes, crushes them, and measures the sugar content. Identify the population, the sample, parameter, and statistic for this situation.

5. An e-commerce company performs a web-based survey by posting a poll on its web site. Since the survey is voluntary, not all people who visit the web site fill out the survey. Identify the population, the sample, parameter, and statistic for this situation.

6. Identify the type of sample obtained (judgment, voluntary, convenience, probability, simple random, stratified, systematic, or cluster).
   (a) A pollster talks to every household in a 4 block region. Then travels to another city and and again talks to every household in a 4 block region.
   (b) A sociologist wants to know the number of hours students at her college spend on the Internet. She knocks on every tenth door in the campus' dorms and interviews one student from each room.

Answers to sample exams.

True/False

1. T 2. T 3. F 4. F 5. F 6. T 7. T 8. F 9. F 10. T 11. T 12. F 13. F
14. F 15. T 16. T 17. F 18. T 19. F 20. F 21. F 22. T 23. T 24. F 25. T 26. T
27. T 28. F 29. T 30. F

Multiple Choice

1. d 2. b 3. d 4. b 5. c 6. d 7. e 8. d 9. e 10. e 11. d 12. e 13. d
14. e 15. a 16. c 17. d 18. a 19. b 20. c

Problems and Open-Ended

1. Statistical inference is the process of arriving at a conclusion about a population parameter on the basis of information obtained from a sample statistic
2. Stratified sample is usually performed when the population is divided into relatively homogeneous groups and there's a large variation within the groups of the population.
3. Population: all customers of their service. Sample: the 500 customers they call. Parameter: the proportion of their customers who want this shopping channel added to their basic cable. Statistic: the proportion of the 500 customers they call who want this shopping channel added to their basic cable.
4. Population: all bunches of grapes in their vineyard. Sample: the 10 bunches they pick. Parameter: the sugar content of the grapes in the vineyard. Statistic: the sugar content of the grapes in the sample.
5. Population: all people who visit their web site. Sample: those who chose to take the survey. Parameter: the population's true values of the question. Statistic: the proportion of the 500 customers they call who want this shopping channel added to their basic cable.
6. (a) This is a cluster sample, the 4-block region is the cluster.
   (b) This is a systematic sample.

T 1. Many statistical techniques have been proven theoretically, but these techniques can still be misapplied and misinterpreted.

T 2. Statistics may be used to mislead without actually lying.

F 3. Bias can only occur in statistics when a conscious effort is made to distort the analysis.

T 4. Two different groups of numbers may have arithmetic means of the same value.

F 5. The arithmetic mean is always the same as a median; that is, the arithmetic mean represents the middle value in a group of numbers.

T 6. Two different groups of numbers may have the same value for an arithmetic mean, but it's possible at the same time that they can have different degrees of dispersion.

F 7. Since numbers such as an average may be used to mislead, it's better to have the data presented graphically because it's not possible to distort visual presentations.

F 8. The values represented on the horizontal axis of a line chart represent causes of the values represented on the vertical axis.

F 9. If a data set is bimodal, then the arithmetic mean is the average of the two modes.

T 10. Statistics may be used to infer relationships but one should never go so far as to state causal relationships from statistics alone.

F 11. If two people are not lying, then they both must have a similar interpretation of a given statistic.

T 12. Statistics may be misleading not because of extraneous confusing summary measures but because of missing information.

F 13. The median value must always equal the mode value.

F 14. The mean is the most commonly occurring value in a group of numbers.

T 15. We may take a wavy line represented on a graph and give an appearance of a straight line by simply changing the proportions of the vertical and horizontal axes.

F 16. In using past trends to project into the future, it is assumed that conditions which have contributed to persistent patterns are constantly changing.

T 17. A simple statement about percentage change does not give us any idea about absolute change.

F 18. If event B follows event A, we can confidently say that A causes B.

T 19. Two events may be shown to have a statistical relationship, but it's possible the relationship is only coincidental.

T 20. An index number is a measure that shows how a composite group has changed with respect to a base period.

F 21. In analyzing data, a statistician should be concerned only with statistical results and not with the selection of methodologies.

F 22. Percentage decreases can easily exceed 100 percent when the original data are positive values.

T 23. Data showing the relationships, changes, and trends contained in alphanumeric reports can often be highlighted with a few graphic presentations.

F 24. A bar chart is one in which data points on a grid are connected by a continuous line to convey information.

T 25. Multiple data sets can be shown on a single line chart.

F 26. Component-part line charts may be prepared, but component-part bar charts aren't feasible.

T 27. Statistical maps present data on a geographical basis.

T 28. A computer graphics package is typically used to convert numeric data into visual images.

F 29. Presentation packages are used to communicate messages to an audience but they cannot be used to analyze data.

1. Bias may occur in statistics when:
   (a) data are intentionally manipulated to produce desired results.
   (b) the input data are incorrect.
   (c) the researcher unintentionally uses data which are not applicable to his problem.
   (d) data are used for purposes other than originally intended.
X (e) All of the above factors are present.

2. Which of the following terms refer to the most commonly occurring value in a group of numbers?
   (a) Arithmetic mean.
X (b) Mode.
   (c) Median.
   (d) Average
   (e) All of the above.

3. In order to describe the spread of a group of values, we should compute the
   (a) arithmetic mean.
   (b) mode.
   (c) median.
   (d) average.
X (e) None of the above.

4. Which of the following may be a measure of central tendency?
   (a) Average.
   (b) Mean.
   (e) Mode.
   (d) Median.
X (e) All of the above.

5. Which of the following presentation methods cannot be distorted to mislead the leader?
   (a) Bar charts.
   (b) Pie charts.
   (e) Graphs.
   (d) Statistical tables
X (e) None of the above.

6. A dispersion measure is an indicator of
   (a) the value of the mean.
   (b) the size of the mode.
   (e) the tendency toward the median.
X (d) the scatter of values about the central measure.
   (e) None of the above.

7. The mode, median and mean for the values in Group A are equal, and the mode, median and mean for the values in Group B are equal. If the mean of Group A is equal to the mean of Group B, how might the two groups differ?
   (a) In the central tendency measures.
   (b) In the most commonly occurring values in each group.
X (c) In the spread of values about the central tendency measure.
   (d) In the averages of each group.
   (e) None of the above.

8. Assume you have been presented a report full of quantitative information. Which of the following should be of concern to you?
   (a) Possible bias by the source of information.
   (b) Methods of data collection.
   (c) Validity of the conclusions.
   (d) Definition of concepts.
X (e) All of the above should be of concern.

9. If two persons receive a report containing statistical results, which of the following is possible?
   (a) Both persons will have the same conclusions.
   (b) Both persons will have different conclusions.
   (c) Both persons may make opposing conclusions without actually lying.
   (d) Both persons may make similar conclusions using different portions of the report.
X (e) All of the above are possible.

10. Which of the following may be used for description of data?
   (a) Statistical tables.
   (b) Measures of central tendency
   (c) Line charts.
   (d) Pictographs.
X (e) All of the above.

11. Computer-generated graphics
   (a) are of little value in presenting statistical data.
X (b) can be prepared by software packages running on personal computers.
   (c) require the resources of large computer systems.
   (d) cannot be produced in color.
   (e) are limited to pie and bar charts

12. A line chart
    (a) is limited to the use of time-series data.
X   (b) doesn't present specific data as well as a table.
    (c) uses the length of a bar to represent a quantity.
    (d) cannot depict multiple data sets.
    (e) has a vertical axis that is usually measured in units of time.

13. A bar (or column) chart
    (a) uses a continuous line to convey information.
    (b) cannot depict percentages.
X   (c) may show how items of interest change over time.
    (d) cannot show negative values.
    (e) cannot be used in combination with other graphic chart forms.

14. Computer graphics packages
    (a) are computer programs that convert pictures into the numeric data that people prefer to use.
    (b) are computer programs that are used to draw pictures.
    (c) are seldom used to analyze data.
    (d) are seldom used for analysis and presentation purposes.
X   (e) can support the use of maps.

15. A presentation package
X   (a) can produce multiple three-dimensional images.
    (b) has few of the features found in an analysis package.
    (c) cannot produce bar or pie charts.
    (d) doesn't have the animation features found in an analysis package.
    (e) doesn't have the color capabilities of an analysis package

1. What is bias?

2. What are six ways in which statistics may be misused.

3. Identify and define three measures of "average".

4. Discuss how dispersion can effect averages.

5. What questions should you ask yourself when you evaluate quantitative information to reduce your chances of being misled?

Answers to sample exams.

True/False

1. T 2. T 3. F 4. T 5. F 6. T 7. F 8. F 9. F 10. T 11. F 12. T 13. F
14. F 15. T 16. F 17. T 18. F 19. T 20. T 21. F 22. F 23. T 24. F 25. T 26. F
27. T 28. T 29. F

Multiple Choice

1. e 2. b 3. e 4. e 5. e 6. d 7. c 8. e 9. e 10. e 11. b 12. b 13. c
14. e 15. a

Problems and Open Ended

1. Bias is the inclination that hampers impartial judgment, and the use of poorly worded and/or biased questions during data gathering may lead to worthless results.
2. (1) Misleading graphs and charts where only a portion of the vertical scale is shown to emphasize to a difference or the horizontal scale is changed to show things in a different light. (2) Use of the converse, since *B follows A*, therefore, *B* was *causes A*. (3) Antics with semantics: failing to define terms that are important; using an alleged statistical fact to jump to a conclusion that ignores other possibilities; and using jargon to cloud the message. (4) Because a pattern has developed in the past in a category, that pattern will continue into the future. (5) Fail to clarify the base period used in computing percentages. (6) Spurious accuracy, precision greater than the data allows.
3. Arithmetic mean or mean is the sum of the values divided by the number of values. Median is the value of the middle position after all the values are arranged in an ascending or descending order. Mode is the score that occurs most often.
4. Dispersion is the amount os spread or scatter that occurs in the data. Many distributions can have the same "average" yet be very different.
5. Who is the source of the information? What evidence is offered by the source in support of the information? What information is missing? And is the conclusion reasonable?

F 1. In the study of statistics, the term "variable" refers to the specific value of a characteristic.

T 2. The number of students attending a philosophy lecture on any given day is a discrete variable.

T 3. If we take raw data, arrange them in ascending order and then group them into some classes, we would be constructing a frequency distribution.

T 4. One of the disadvantages of a frequency distribution is the loss of detail in the data.

F 5. A histogram is used only to represent a cumulative frequency distribution.

T 6. In computing the mean for the grouped data, it is assumed that the values within a class are concentrated about the class midpoint.

T 7. Open-ended classes should be avoided because important descriptive measures for the distribution such as the arithmetic mean and the standard deviation are impossible to compute.

F 8. An interquartile range includes approximately the upper 50 percent of the values.

T 9. The extreme values in a positively skewed distribution are in the right tail.

T 10. In a symmetrical distribution, the values of the mean, median and mode are the same.

F 11. Assume we have two distributions which are both symmetrical, have the same mean value and the same standard deviation value. We can consequently say that these two distributions are exactly the same.

T 12. The width of each class in a frequency distribution is called the class interval.

F 13. Extremely high or low values don't distort the mean.

T 14. No single point on an ogive can exceed 100 percent.

F 15. If there are any overestimates or underestimates in each class based on the assumption that each observation in a class has a value equal to the class midpoint, the overall mean cannot be computed because all these errors accumulate to sizable error.

F 16. The use of the quartile deviation is seldom appropriate in badly skewed distributions.

T 17. The method of computing an overall mean for grouped data shown in the text ignores the width of class intervals.

T 18. A median is a better measure of central tendency than the mean when a distribution is badly skewed.

T 19. Of the three measures of central tendency, only the mean cannot be determined with an open-ended distribution.

T 20. The sum of the differences between the data items and the mean is zero.

T 21. With a skewed distribution, the mean is different from the mode and the mean is in the direction of the tail of the distribution.

F 22. In a skewed distribution, the only measure of central tendency which remains under the peak of the curve is the mean.

T 23. There is only one way for a set of numbers to have standard deviation of zero and that is to have identical numbers.

F 24. Based on statistical theory, an arithmetic mean is the most representative descriptive measure of dispersion.

F 25. It can be mathematically proven that two sets of numbers cannot produce the same average mean.

T 26. Two sets of numbers may possess the same mean, but they may also possess different degrees of dispersion.

T 27. An arithmetic mean may be representative of the data being studied if the data values are relatively uniform.

T 28. The variance is the standard deviation squared.

F 29. A standard deviation is a measure of relative dispersion.

T 30. The standard deviation may be used with the mean to indicate the proportions of observations in the distribution that fall within specified distances from the mean.

F 31. A measure of dispersion gives some idea about the size of the average mean.

T 32. Measures of absolute dispersion are expressed in units of the original observation.

F 33. To compute the mean absolute deviation, we square all deviations to eliminate negative values.

T 34. A stem-and-leaf display may be used in exploratory data analysis.

F 35. A stem-and-leaf display resembles a frequency polygon rather than a histogram.

F 36. There are usually as many stems in a stem-and-leaf display as there are items in a data set.

F 37. Although we may have a large mass of data, statistical techniques allow us to adequately describe and summarize the data with an average.

T 38. A skewed distribution occurs when a few values are much larger or smaller than the typical values found in a data set.

T 39. The mean of the squared deviations about the mean is known as the variance.

T 40. The variance is not expressed in the units of the original data.

T 41. Values are often found to be distributed symmetrically about their mean.

T 42. The quartile deviation is similar to the range in that both are measures of distance between two selected points.

T 43. The smaller the quartile deviation, the less dispersion from the mean there will be in the middle half of the observations in the distribution.

1. Which of the following is an example of an attribute?
   (a) Number of people in line.
   (b) Weight of an orange.
   (c) Length of a loaf of bread.
X (d) Color of the grapes on sale.
   (e) The temperature of a frozen turkey.

2. Given the following data, what is the mean, median and mode?

   82 98 73 71 82 90 43

X (a) mean = 77; median = 82; mode = 82.
   (b) mean = 82; median = 80; mode = 82.
   (c) mean = 80: median = 86; mode = 80.
   (d) mean = 84; median = 84; mode = 84.
   (e) None of the above.

3. Assume we have a set of data which have values ranging from 94 to 172. If we wish to construct a frequency distribution what would be the size of the intervals?
   (a) 6.
   (b) 10.
   (c) 12.
   (d) 13.
X (e) Cannot be computed without more information.

4. If we have a set of data with a high value of 160 and a low value of 96, and if we wish to construct a frequency distribution in ascending order with eight classes, what would be the first interval listed?
   (a) 96 and less.
X (b) 96 and less than 104.
   (c) 152 and less than 160.
   (d) 160 and over.
   (e) Cannot be computed without more information.

5. What type of distribution is described by the following information?

   mean = 6.5; median = 6.3; mode = 5.4.

   (a) Negatively skewed.
   (b) Symmetrical.
   (c) Bimodal.
X (d) Positively skewed.
   (e) Distribution unknown.

6. Which of the following cannot be computed if we have an open ended frequency distribution?
   (a) median.
X (b) mean.
   (c) mode.
   (d) median and mode.
   (e) mean and median.

7. If we have a set of data with a high value of 243 and a low value of 194, and if we wish to construct a descending frequency distribution with 7 classes, what would be the second interval listed?
   (a) 194 and less than 201.
   (b) 201 and less than 208.
X (c) 229 and less than 236.
   (d) 236 and less than 243.
   (e) None of the above.

8. What type of distribution is described by the following information?

   mean = 46; median = 48.1; mode = 53.

X (a) Negatively skewed.
   (b) Symmetrical.
   (c) Bimodal.
   (d) Positively skewed.
   (e) Distribution unknown.

9. What type of distribution is described by the following information?

   mean = 164; median = 142; mode = 173.

   (a) Negatively skewed.
   (b) Symmetrical.
   (c) Bimodal.
   (d) Positively skewed.
X (e) Distribution unknown.

10. The following frequency distribution shows the time (in seconds) that a group of 20 people could hold their breath under water.

Seconds	Number of Persons
5 to less than 10	3
10 to less than 15	6
15 to less than 20	9
20 to less than 25	2

Therefore, the mean is
   (a) 12.5.
   (b) 14.0.
X (c) 15.0.
   (d) 16.0
   (e) 17.5.

11. Which of the following statements is true:
X (a) A discrete variable has a countable or limited number of values.
(b) A continuous variable is counted rather than measured.
(c) An array must place values in an ascending order.
(d) Arrays always place values in a descending order.
(e) Average of nominal measures are useful values.

12. Consider the following frequency distribution of the number of classical records sold per day in a particular 40 day period.

Records Sold/Day	Number of Days
1 to 5	0
6 to 10	15
11 to 15	18
16 to 20	7

Therefore, the mean is
(a) 7.4.
(b) 9.3.
(c) 10.0.
(d) 11.0.
X (e) None of the above.

13. Determine the median of the following frequency distribution:

Classes	Frequency
14 to 18	4
19 to 23	16
24 to 28	2
29 to 33	18
34 to 38	10

(a) 25.39.
(b) 26.0.
(c) 29.39.
X (d) 29.83.
(e) 32.17.

14. Which of the following statements is true?
(a) The histogram is a line chart of a frequency distribution.
(b) An ogive is most often used to graphically locate the mode.
X (c) An ogive is a graph of a cumulative frequency distribution.
(d) Whenever possible, the width of classes in a frequency distribution should be unequal.
(e) A frequency polygon is a bar chart of a frequency distribution.

15. A stem-and-leaf display
    (a) places a line at the median value of the data.
    (b) has a leaf value that represents the leading digit(s) of all data items listed on a row.
X (c) is a plot that looks like a histogram, but it includes the actual data values.
    (d) is a plot that looks like a frequency polygon, but it includes actual data values.
    (e) has as many stems as there are data items in a data set.

16. Although a measure of central tendency is useful in describing and summarizing data, which of the following is also needed to describe and summarize the data?
    (a) Modal value.
    (b) Algebraic mean.
    (c) Measure of the median.
X (d) Dispersion measure.
    (e) All of the above.

17. Which of the following is true?
    (a) A relatively uniform set of numbers will have little dispersion.
    (b) A measure of central tendency and a measure of variability are needed to adequately describe a set of data.
    (c) Two different sets of data may produce the same arithmetic mean.
    (d) A measure of dispersion is an indicator of the reliability of the average value.
X (e) All of the above are true.

18. Which of the following is the crudest measure of dispersion?
    (a) Median.
    (b) Mode.
    (c) Absolute dispersion.
X (d) Range.
    (e) Mean absolute deviation.

19. Which of the following is correct?
    (a) Two sets of numbers with completely different means and different mean absolute deviations may have the same range.
    (b) The most widely used measure of dispersion is the standard deviation.
    (c) The mean absolute deviation differs from the standard deviation because the mean absolute deviation ignores the algebraic signs.
    (d) The range is a crude measure of dispersion because it does not consider all observations.
X (e) All of the above are correct.

20. The standard deviation will always be larger than the mean absolute deviation for the same data because:
    (a) the standard deviation considers all the data and the mean absolute deviation uses only data which are less than the mean.
X   (b) the extreme values receive stronger emphasis when they are squared.
    (c) the standard deviation uses the extreme values while the mean absolute deviation does not use them.
    (d) the extreme values are used to compute a range and only the standard deviation uses these range of values.
    (e) None of the above.

21. Which of the following is not true?
    (a) The interquartile range is computed with the first and third quartile positions.
X   (b) The interquartile range cannot be computed in an open-ended distribution.
    (c) The second quartile position is the median.
    (d) The lower 25 percent of the values are not considered in the computation of the quartile deviation.
    (e) The interquartile range is expressed in the original units of observation.

22. Which of the following is true about the range?
    (a) It is an easy measure to compute.
    (b) The range is expressed in the original units of observation.
    (c) Its computation is based only on the extreme values.
    (d) It ignores representative items of the distribution.
X   (e) All of the above are true.

23. Which, if any, of the following is not a characteristic of the standard deviation?
X   (a) It is the most popular measure of central tendency.
    (b) It is affected by the value of every observation.
    (c) It is distorted by extreme values.
    (d) It can't be found in an open-ended distribution.
    (e) All of the above are standard deviation characteristics.

24. What is the variance of the following numbers?

    25 35 15 20 40

    (a) 5.4
    (b) 9.27
    (c) 74.2
X   (d) 86.0
    (e) 88.4

25. Assume we have the following distribution of grades for a statistics exam for 10 students:

Grade Intervals	Number of Students
90 to less than 100	5
80 to less than 90	3
70 to less than 80	2

Which of the following is correct?

(a) Mean = 84; Variance = 43.
(b) Mean = 78; Variance = 64.
(c) Mean = 85; Variance = 10.
X (d) Mean = 88; Variance = 61.
(e) Mean = 89; Variance = 5.

1. Why is it necessary to measure dispersion to describe a data set.

2. Describe three measures of dispersion.

3. What is a variable?

4. What five values are needed to construct a box-and-whiskers display?

5. Discuss the advantages and disadvantages of Chebyshev's theorem.

6. Discuss the differences between qualitative and quantitative. What types of statistics can only be found with quantitative data?

7. What is the interquartile range?

8. Calculate the mean and median for the following data.
   82 98 73 71 82 90 43

9. Construct a box-and-whiskers display for the following data.
   52 55 57 57 61 63 64 64 65

Answers to sample exams.

True/False

1. F 2. T 3. T 4. T 5. F 6. T 7. T 8. F 9. T 10. T 11. F 12. T 13. F
14. T 15. F 16. F 17. T 18. T 19. T 20. T 21. T 22. F 23. T 24. F 25. F 26. T
27. T 28. T 29. F 30. T 31. F 32. T 33. F 34. T 35. F 36. F 37. F 38. T 39. T
40. T 41. T 42. T 43. T

Multiple Choice

1. d 2. a 3. e 4. b 5. d 6. b 7. c 8. a 9. e 10. c 11. a 12. e 13. d
14. c 15. e 16. d 17. e 18. d 19. e 20. b 21. b 22. e 23. a 24. d 25. d

Problems and Open Ended

1. Averages alone don't adequately describe a data set. Two data sets can have the same mean and median and be quite different. Dispersion tells how well the average value depicts the data. Also to learn the extent of the scatter so that step may be taken to control the existing variation.
2. Range, mean absolute deviation, standard deviation.
3. Aa variable is a characteristic of interest that is possessed by each item under study.
4. Minimum, Q_1, median, Q_3, and maximum.
5. Advantage: Chebyshev's theorem can be used on any distribution. Disadvantage: We can do much better than Chebyshev's theorem on bell shape distributions (and other distributions we will study later).
6. Qualitative data are names that reveal nothing about size or rank. Quantitative data consist of variables that are counted or measured. With quantitative data we can find the mean and standard deviation. With qualitative we can only the relative frequency with which a particular characteristic occurs.
7. The interquartile range includes approximately the middle 50 percent of the values. It is the difference between the third quartile, Q_1, and the first quartile, Q_3.
8. Mean = 77 and median = 82.
9. Minimum $= 51$, $Q_1 = 56$, median $= 61$, $Q_3 = 64$, and maximum $= 65$.

T 1. The probability of an event refers to the chance or likelihood that an event will occur.

F 2. The probability of an event's occurrence must always be determined by observation and experimentation.

F 3. A sample space is any subset of the possible simple outcomes, responses, or measurements of an experiment.

T 4. Although the probability of an event occurring is 9/10, the event may not occur at all in 10 trials.

F 5. If the probability of a particular event A is 1/3, and if after two trials event A did not occur, event A must occur on the third trial.

F 6. If the probability of a particular event is 24 percent, then the event cannot occur more than once in four trials.

T 7. An a priori probability can be determined without observation over a large number of trials.

F 8. If an experiment has 5 possible outcomes, then the probability of each outcome is 1/5.

T 9. If we have two mutually exclusive events, the probability of either one of these events occurring may be determined by summing the probabilities of each of these events.

T 10. If $P(A \text{ or } B) = P(A) + P(B)$, then the events A and B are mutually exclusive.

F 11. If $P(A|B) \neq P(A)$, then the events A and B are mutually exclusive.

T 12. If two events are not mutually exclusive, there is a possibility that both events may occur at the same time.

T 13. If two events are independent, the occurrence of one event should not affect the likelihood of occurrence for the other event.

T 14. If $P(A|B) = P(A)$, then the events A and B are independent.

T 15. Assume there are two events A and B which are not mutually exclusive. In order to determine the probability of either A or B occurring, we must add the individual probabilities for each event and then subtract their joint probability.

F 16. If $P(A) + P(B) > 1$, then the events A and B are independent.

T 17. If a joint probability for two events is to be computed it is necessary to know if the two events are independent.

F 18. If a machine produces 18 defective parts out of a total of 200 parts produced, the percentage of defective parts is 18.

F 19. If two events are independent, the computation of their joint probability requires knowledge of the conditional probabilities rather than the individual probabilities.

F 20. If events A and B are independent and P(A) = 0.5 and P(B) = 0.5, their joint probability would be equal to one.

T 21. If we have four items--A, B, C, and D, there would be six different pairs of items.

F 22. If we have five items--A, B, C, D, and E, the probability of selecting the pair AE out of all possible pairs is 5 percent.

T 23. The probability distribution of a variable which has a finite number of values is a discrete distribution.

F 24. If events A and B are mutually exclusive and P(A) = 0.5 and P(B) = 0.2, then P(A or B) = $0.5 \cdot 0.2 = 0.010$.

F 25. A subjective probability is one that's determined from observation and experimentation.

T 26. An expected value is a weighted mean of all the possible values that a discrete random variable can assume in an experiment.

F 27. The expected value is always positive.

T 28. A joint probability is the probability that both events will occur.

T 29. A random variable is one that takes on different values depending on the chance outcome of an experiment.

T 30. The complement of an event E consists of all possible events in the sample space that are not in event E.

T 31. A random variable has a single numerical value for each outcome of a probability experiment.

F 32. The only rule that applies to all probability distributions is that the possible probability distribution values are always between 0 and 1.

1. Which of the following statements is false?
   (a) Probability may be expressed as a number between zero and one.
   (b) A probability of 0.25 for a particular event does not guarantee that the event will occur once in four trials.
   (c) An a priori probability may be revised after empirical research.
X (d) If the occurrence of one event affects the likelihood that another event will occur, we have two independent events.
   (e) Mutually exclusive probability situations involve addition.

2. Assume event A and event B are independent. If the probability of event A occurring is 60 percent, and the probability of event B occurring is 20 percent, which of the following is true?
X (a) P(A and B) = 0.12
   (b) P(A or B) = 0.12
   (c) P(A or B) = 0.32
   (d) P(A or B) = P(A) + P(B) – P(A and B) = 0.4
   (e) P(A and B) = 0.8

3. There are two baseball leagues which do not engagc in any intcrlcaguc games during the playing season. Each league sends its top team to the Universe Series at the end of the regular season. Team A in one league has a 30 percent chance of finishing first while Team Z in the other league has a 40 percent chance of playing in the Universe Series. What are the chances of Team A being in the Universe Series and Team Z not being in the Series?
   (a) 12 percent
X (b) 18 percent
   (c) 24 percent
   (d) 58 percent
   (e) 70 percent

4. A deck of cards has four suits; that is, hearts, diamonds, clubs, and spades. Each suit has an ace, king, queen, jack, and the numbers two through ten. If the deck is well shuffled, what is the probability that a king of spades or any club card will be drawn?
X (a) 7/26
   (b) 2/13
   (c) 3/52
   (d) 4/13
   (e) 2/9

5. Assume that events A and B are mutually exclusive where P(A) = 0.6 and P(B) = 0.3. Which of the following statement is false?

X (a) P(A and B) = 0.18
(b) P(A compliment) = 0.4
(c) P(B compliment) = 0.7
(d) P(A or B) = 0.9
(e) None of above.

6. Assume there are 15 identical envelopes. Five of these envelopes each contain $1. The remaining 10 envelopes contain $20 in each envelope. If all the envelopes were placed in a box and mixed well, what is the probability that an individual will randomly select an envelope with $20 on the first pick and then after replacing the envelope he will randomly select a $1 envelope?

X (a) 2/9
(b) 3/7
(c) 1/3
(d) 1/2
(e) None of the above.

7. Assume you and a friend are among 25 persons eligible for prizes in a lottery. The prize for being selected first is $10,000 while the second prize is $7,500. Assuming that a person will be selected randomly and may win only one prize, what are the chances that you and your friend together win $17,500, irrespective of which one is first or second?

(a) 2/25
X (b) 1/300
(c) 2/625
(d) 1/49
(e) 1/12

8. Assume that events A and B are independent where P(A) = 0.6 and P(B) = 0.3. Which of the following statement is false?

(a) P(A and B) = 0.18
(b) P(A compliment) = 0.4
(c) P(B compliment) = 0.7
X (d) P(A or B) = 0.9
(e) None of above.

9. A probability experiment has 3 possible outcomes, A, B, and C. If P(A) = 0.6 and P(B) = 0.2, then P(C) = ____.

(a) 0.12
X (b) 0.2
(c) 0.4
(d) 0.8
(e) None of above.

10. Assume that $P(A) = 0.7$ and $P(B) = 0.6$. Which of the following statement is always true?
    (a) Events A and B are independent.
    (b) Events A and B are mutually exclusive.
    (c) Events A and B are not independent.
X   (d) Events A and B are not mutually exclusive.
    (e) None of above.

1. Define sample space and a simple event.

2. Explain the concepts of independent and mutually exclusive events. Can two events be both independent and mutually exclusive.

3. What is a random variable?

4. What is the measure of central tendency for a random variable called and how is it computed.

5. What are the two rules that apply to any probability distribution?

6. How are discrete random variables different from continuous random variables?

7. What is conditional probability?

Answers to sample exams.

True/False

1. T 2. F 3. F 4. T 5. F 6. F 7. T 8. F 9. T 10. T 11. F 12. T 13. T
14. T 15. T 16. F 17. T 18. F 19. F 20. F 21. T 22. F 23. T 24. F 25. F 26. T
27. F 28. T 29. T 30. T 31. T 32. F

Multiple Choice

1. d 2. a 3. b 4. a 5. a 6. a 7. b 8. d 9. b 10. d

Problems and Open Ended

1. A sample space is the set of all possible simple outcomes, responses,or measurements of an experiment. A simple event is one that can't be broken down any further.
2. Two events are independent when the occurrence (or nonoccurrence) of one does not affect the probability of the occurrence of the other. Two events are mutually exclusive if the occurrence of one during a single trail of an experiment prevents the occurrence of the other during the same trail, that is, the two events have no simple events in common. Two sets cannot be both independent and mutually exclusive.
3. A random variable has a single numerical value for each outcome of a probability experiment.
4. The measure of central tendency for a random variable is called the expected value. It is computed as the sum of the random variable this the probability that that random variable. That is $E = \Sigma[x \cdot P(x)]$, where: $E =$ the expected value; $x =$ the value of each possible outcome; $P(x) =$ the probability of that possible outcome.
5. (1) The values of the probability are numbers between 0 and 1. That is, $0 \leq P(x) \leq 1$. (2) All of the possible values in a probability distribution must add up to 1. That is, $\Sigma P(x) = 1$.
6. A discrete random variable is one in which all possible values can be listed while continuous random variables can take on an infinite number of number of values.
7. The probability that event B occurs, given that event A has already happened.

F 1. A probability distribution is simply a listing of all possible outcomes of an experiment.

T 2. ${}_4C_4$ is equal to ${}_4C_0$.

F 3. In a binomial experiment the probability of success is always greater than or equal to probability of failure.

T 4. In a binomial experiment each trial is independent of the other trials.

F 5. ${}_6C_4$ is equal to ${}_6C_3$.

T 6. A binomial distribution is a discrete probability distribution.

F 7. In a binomial distribution the standard deviation is npq, where n is the number of trials, p is the probability of success, and q is the probability of failure.

T 8. A binomial distribution describes the distribution of probabilities when there are only two possible outcomes on each trial of an experiment.

T 9. In a Poisson distribution the variance is equal to the mean.

F 10. The Poisson distribution requires that the average number of occurrences of an event vary for each unit of time or space.

F 11. The Poisson distribution is a continuous probability distribution.

T 12. The Poisson distribution requires that the number of occurrences of an event in one unit of time or space is independent of the number that occur in other such units of time or space.

T 13. With a normally distributed set of values, virtually all of the values would fall within three standard deviations of the mean.

T 14. An important characteristic of the normal distribution is that 50 percent of the area under the curve falls to the left of the mean while the other 50 percent of the area under the curve falls to the right of the mean.

T 15. If we have two different normal curves, one standard deviation from the mean in each of these curves covers the same area.

T 16. Since the normal curve is a probability distribution, the total area under the curve is one.

F 17. In the standard normal distribution the mean is always 0 while the standard deviation may vary.

F 18. A Poisson distribution is symmetric about its mean.

T 19. A continuous probability distribution allows the variable to be measured to any degree of precision.

T 20. Probabilities for continuous distributions are represented by areas under the curve.

1. Buck Shott, an expert with a rifle, hits a bulls eye 90 percent of the time when shooting. What is the probability of Buck hitting exactly 2 bulls eyes in 4 shots?
   (a) 0.0063
   (b) 0.0324
X (c) 0.0486
   (d) 0.09
   (e) 0.81

2. A student has been able to correctly answer true-false questions in psychology quizzes 80 percent of the time. If he has a psychology quiz with 5 true-false questions, what is the probability that he will correctly answer exactly 3 questions?
   (a) 0.0248
   (b) 0.1600
X (c) 0.2048
   (d) 0.5121
   (e) 0.7848

3. Assume there are 6 students in a seminar in Sociology. There is a unique project to be done by two students. The instructor has written down all possible pairs of students and will randomly select the pair. Student A and Student Z were wondering what were their chances of being selected. The probability of selecting this pair is:
   (a) 1/6
   (b) 1/12
X (c) 1/15
   (d) 1/30
   (e) 1/36

4. Assume there are two lakes containing trout and the weights of the trout in each lake are normally distributed. The distribution of trout weight in Lake A has a mean of 18 ounces with a standard deviation of 2 ounces. The distribution of trout weight in Lake B has a mean of 19 ounces and a standard deviation of 3 ounces. Which of the following statements is not true?
   (a) The chances of catching a trout 18 ounces or less is greater in Lake A than in Lake B.
   (b) The chances of catching a trout 18 ounces or more in Lake A is the same as the chances of catching a trout 19 ounces or more in Lake B.
   (c) The chances of catching a trout between 18 and 20 ounces in Lake A are the same as the chances of catching a trout between 16 and 19 ounces in Lake B.
   (d) The chances of catching a trout weighing 18 to 20 ounces in Lake A are the same as the chances of catching a trout between 16 and 18 ounces in the same lake.
X (e) A person has the same chance of catching a trout between 15 and 17 ounces in Lake A as catching a trout between 16 and 18 ounces in Lake B.

5. A baseball player gets a "hit" 30 percent of the time he is at bat. Assume that the player will be at bat four times in an upcoming game. What is the probability that he will get exactly 3 hits during the game?
   (a) 0.0133
   (b) 0.021
   (c) 0.027
X (d) 0.0756
   (e) 0.27

6. If we have a set of normally distributed values with a mean of 88 and a standard deviation of 6, within which range would we expect approximately 95 percent of the values to fall?
   (a) 70 to 106
X (b) 76 to 100
   (c) 82 to 94
   (d) 88 to 94
   (e) 88 to 106

7. The weight of pinochle players in a certain league is normally distributed with a mean of 144 pounds and a standard deviation of 6 pounds. The chances of a player weighing between 144 and 156 pounds are
X (a) 0.4772
   (b) 0.4886
   (c) 0.5000
   (d) 0.5248
   (e) 0.5434

8. The lifetimes of batteries produced by a firm are normally distributed with a mean of 100 hours and a standard deviation of 10 hours. What is the probability a battery will last between 110 and 120 hours?
   (a) 0.0680
X (b) 0.1359
   (c) 0.1587
   (d) 0.2946
   (e) 0.8413

9. The grades for Economics students at a large university are normally distributed with a mean of 76 and a standard deviation of 4. What is the probability that a randomly selected student will have a grade between 84 and 88?
X (a) 0.0215
   (b) 0.0430
   (c) 0.3413
   (d) 0.4772
   (e) 0.4987

10. Assume we have a bag containing 4 red marbles and 6 green marbles. A person randomly selected one marble and then without replacement of the first marble randomly selected a second marble. The probability of selecting 2 red marbles is:
    X (a) 0.1333.
    (b) 0.1667.
    (c) 0.2443.
    (d) 0.6718.
    (e) 0.76.

11. Which of the following condition is not characteristic of a Poisson distribution?
    (a) The distribution is discrete.
    (b) The probability that an event occurs is the same for each unit of time or space.
    (c) An experiment consists of counting the number of times a certain event occurs during a given unit of time or space.
    (d) The number of events that occur in one unit of time or space is independent of the number that occurs in other such unit.
    X (e) All of the above are correct.

12. Telephone calls at a police station follow a Poisson process and occur at an average rate of six calls per hour. The probability that there will be exactly four calls in an hour is:
    X (a) 0.1339.
    (b) 0.1647.
    (c) 0.3219.
    (d) 0.6558.
    (e) None of the above.

13. Using the data from question 12 above, the probability that there will be at least four calls in an hour is:
    (a) 0.1512.
    (b) 0.6269.
    X (c) 0.8488.
    (d) 0.9323.
    (e) None of the above.

14. In probability terminology,
    (a) two events are said to be independent if the occurrence of one prevents the occurrence of the other.
    (b) a joint probability is the probability that only one of two events will occur.
    X (c) two events are said to be dependent if the probability of occurrence of one is conditioned by whether or not the other happens.
    (d) a random variable has the same value regardless of the chance outcome of an experiment.
    (e) a probability distribution is a listing of one possible outcome of an experiment.

15. A discrete probability distribution
    (a) was first created by Simeon Denis Discrete, the Frenchman who developed it in the early 1800s.
X   (b) has a finite number of values that a random variable can take.
    (c) consists of continuous random variables with an infinite number of values.
    (d) is the term used to describe the normal distribution.
    (e) is required when variables such as weight, length, or time are considered.

1. What are the properties of a binomial experiment?

2. What are the characteristics of a Poisson distribution?

3. What are the properties of the normal distribution.

4. Which distribution are discrete and which are continuous?

5. What is the standard normal distribution?

6. In a binomial experiment in which the probability of success is 0.75, what is the probability of getting exactly 3 successes in 4 trials?

Answers to sample exams.

True/False

1. F 2. T 3. F 4.T 5. F 6. T 7. F 8. T 9. T 10. F 11. F 12. T 13. T
14. T 15. T 16. T 17. F 18. F 19. T 20. T

Multiple Choice

1. c 2. c 3. c 4. e 5. d 6. b 7. a 8. b 9. a 10. a 11. e 12. a 13. c
14. c 15. b

Problems and Open Ended

1. (1) The same action (trail) is repeated a fixed number of times. (2) Each trail is independent of the others. (3) For each trail, there are just two outcomes of interest. One outcome may be designated "success," and the other "failure." (4) The probability of success remains constant for each trail.
2. In a Poisson distribution we are interested in the number of specific occurrences that take place within a unit of time or space.
3. A normal distribution curve is symmetric and bell shape. The total area under the curve is 1. The probability that a normally distributed random variable will take on a value between a and b is the area under the normal curve between two vertical lines erected at points a and b.
4. Binomial and Poisson are discrete probability distribution and the normal distribution is continuous.
5. It is the specific normal distribution curve that has a mean of 0 and a standard deviation of 1.
6. Probability $= {}_4C_3 \cdot (0.75)^3 \cdot (0.25) = 0.422$.

T 1. A group of items may be a population to one researcher while the same group could be a sample to another.

T 2. A particular measure could be a parameter for one researcher and a statistic for another.

T 3. Sample statistics may be used to make inferences about the population parameters.

F 4. A sample is always preferred because it provides more accurate information than a census.

T 5. A sample may contain a large number of items from a population and still not be representative of the population.

F 6. Any time a sample costs less than a census, we have a sufficient reason for sampling.

T 7. There are situations where a census is possible but where sampling is more feasible.

F 8. In general, an increase in the representativeness of a sample increases the size of the sampling error.

F 9. It is always feasible to study an entire finite population.

T 10. A table of random numbers will generally produce an unbiased sample selection regardless of the manner in which the table numbers are used as long as the researcher picks the numbers in a systematic way.

T 11. If a table of random numbers is used, extremely high values are as likely as extremely low values.

F 12. For the sampling distribution to approximate the normal probability distribution the sample size must be greater than 30.

F 13. If the sample size (n) times the sample percentage (p) is $\geq$ 500, then the sample distribution will approximate a normal probability distribution.

F 14. If we have a population of 100 items and we wanted a simple random sample of 10 items, each item would have a probability of 0.1 of being selected for the sample on the first pick.

T 15. By chance, it's possible that a sample mean will be equal to the population mean.

F 16. The sample mean always underestimates the population mean because of sampling variation.

T 17. Sample distribution can have any shape.

F 18. A sampling distribution of means is the distribution of items in each possible sample combination.

T 19. A sampling distribution and a sample distribution are both frequency distributions.

T 20. The sampling distribution of means is normally distributed when the population is normally distributed.

T 21. If the sampling distribution of means approximates the normal distribution, we may determine the likelihood of a specific outcome by simply knowing the population mean value and the standard error.

T 22. Probability statements can be made concerning the possible mean of a random sample if we know the population mean, the population standard deviation, the sample size, and the population size.

F 23. The standard error of the sample distribution may be computed by simply knowing the population variance and the sample size.

F 24. Any time we deal with a finite population, we must use a finite correction factor in computing the standard deviation of the sampling distribution.

F 25. As the sample size increases the standard deviation of the population decreases, and the range of possible values which a sample mean may assume also decreases.

T 26. Based on the Central Limit Theorem, we can say that a sampling distribution will be approximately normal if the sample size is sufficiently large.

T 27. The mean of a sampling distribution of percentages with simple random samples of size 40 is equal to the population percentage.

T 28. A statistical software package can produce simulations that demonstrate the validity of the Central Limit Theorem.

F 29. Reducing the value of a standard error requires that we reduce sample size.

T 30. A key measure used to summarize the relative frequency with which a particular characteristic occurs is the percentage or proportion.

1. Which of the following does not apply to the term population?
   (a) A population may be infinite in size.
   (b) A population is defined by the researcher.
   (c) A population for one researcher may be a sample for another researcher.
   (d) A population is not used only for people.
X (e) All of the above apply to a population.

2. Which of the following statements is not correct?
   (a) A census is a study of the entire population.
   (b) A parameter is a characteristic of a population.
   (c) A statistic is any characteristic of a sample.
X (d) The value of a statistic can never be more than the value of a parameter.
   (e) The value of a statistic may change with different samples.

3. Which of the following is not correct?
   (a) μ equals the population mean.
   (b) σ equals the population standard deviation.
   (c) N is the total number of items in the population.
   (d) s is the standard deviation of the sample.
X (e) All of the above are correct.

4. Which of the following is not correct?
   (a) Sampling is often needed because a census involves too much time and costs too much.
   (b) Summary measures of a sample provide estimates of population values.
   (c) If a large sample of items is representative of the population, we will generally have reliable estimates of the parameter.
X (d) Usually, a well-designed sample eliminates all error in estimating a parameter.
   (e) Sampling error arises because of the difficulty in obtaining completely representative items from the population.

5. Which of the following is not correct?
   (a) Samples will generally contain error.
   (b) There are usually errors in sampling, but the results of the sample are still useful.
   (c) The precision of the estimate can be objectively assessed by assessing the sampling error.
   (d) Statisticians often feel that the benefits of a census will be too costly, and thus they settle for the advantages to be gained from sampling.
X (e) If a statistician notices the existence of sampling error, his job is to completely eliminate it before estimation.

6. Cost is a ______ reason for sampling.
   (a) sufficient
   (b) irrelevant
X (c) acceptable but not sufficient
   (d) sole
   (e) necessary but unacceptable

7. The ______ of sample results in estimating parameters depends on the sample's representation of a population.
   (a) cost
   (b) speed
X (c) accuracy
   (d) quantity
   (e) None of the above.

8. Which of the following is not true concerning the table of random numbers?
   (a) The occurrence of each digit is determined solely by probability.
   (b) Each digit has a probability of .1 of occurring.
   (c) Each sequence or group of five digits has an equal chance.
   (d) Every digit or every sequence of digits is equally likely if the researcher systematically selects them.
X (e) A particular sequence of digits such as 1 2 3 4 5 should never occur more than once because each possible sequence has an equal chance of occurrence.

9. If there is a population of 10 items, and if we wish to select a simple random sample of 3 items, which of the following is true?
   (a) The probability that a particular combination of 3 items will be selected is 1/10.
   (b) The probability that a particular item or a particular combination of 3 items will be selected is 1/10.
   (c) Each possible combination of 3 items has a 3/120 chance of being selected.
   (d) Each possible combination of 3 items has a 3/10 chance of selection.
X (e) The probability that a particular item will be selected for the sample is 3/10 while each possible combination of 3 items has a 1/120 chance of selection.

10. Which of the following statements is not true concerning a sampling distribution?
   (a) The grand mean of the sampling distribution is equal to the population mean.
   (b) The sampling distribution is a frequency distribution of sample means where the sample sizes are all equal.
   (c) If we have a population of 10 items, the sampling distribution for samples of two items would consist of 45 items.
X (d) The grand mean of the sampling distribution is the arithmetic mean of the distribution divided by the sample size.
   (e) If the sample size is sufficiently large, the sampling distribution approximates the normal distribution.

11. If we have an infinite population with a variance of 36 and a mean of 96 and if a sample of 4 items is taken, what range of values would provide a 68 percent chance of occurrence for a sample mean.
    (a) 94.5 to 97.5
X (b) 93 to 99
    (c) 87 to 105
    (d) 78 to 114
    (e) None of the above.

12. Assume we have a population of 100 items with a mean of 240 and a standard deviation of 24. With a simple random sample size of 16 which of the following is true?
    (a) The mean of the sampling distribution is 80.
    (b) The standard deviation of the sampling distribution is Ö`2416 Ö`100 - 1699 .
    (c) There is a 95 percent chance that the sample mean will be between 232 and 248.
    (d) The mean of the sampling distribution will be 240 and the standard deviation of the sampling distribution will be 6.
X (e) The mean of the sampling distribution equals 240 and the standard deviation of the sampling distribution equals 6 times the finite correction factor.

13. If an infinite population has a mean of 250, a standard deviation of 30 and a simple random sample of 9, which of the following is true?
    (a) The mean of the sampling distribution is unknown based on these data.
    (b) There is approximately a 95 percent chance that the mean of the sampling distribution is between 242.33 and 257.67.
    (c) There is approximately a 68 percent chance that the sample mean will be within 30 of the true mean.
X (d) There is a 95 percent chance that the sample mean will be within approximately 20 of the true mean.
    (e) All of the above are true.

14. A parameter
X (a) is a characteristic of a population.
    (b) is a characteristic of a sample.
    (c) value is called a statistic.
    (d) value is called a sample.
    (e) is identified by lowercase Roman letters.

15. The term "sampling variation" refers to variation in a
    (a) population parameter.
X (b) sample statistic.
    (c) population mean.
    (d) population percentage.
    (e) judgment sample but not a random sample.

16. If a computer simulation produces 100 random samples of the same size from the same population,
    (a) the sample means will be identical.
    (b) the sample standard deviations will be identical.
X (c) the mean of these 100 simulated sample means will likely be close to the population mean.
    (d) the mean of these 100 simulated sample means must equal the population mean.
    (e) the mean of the 100 sample means in one simulation must equal the mean of 100 sample means in a following simulation.

17. A sampling distribution of percentages has
    (a) a standard deviation that's equal to the population mean.
X (b) a mean that's equal to the population percentage.
    (c) a mean that's equal to the population mean.
    (d) a standard deviation that increases as the sample size increases.
    (e) a standard deviation that increases as the population percentage decreases.

1. What are the different type of populations?

2. What are the advantages of sampling?

3. What is sampling distribution of means?

4. Under what conditions does the sampling distribution approximate the normal distribution?

5. How does the size of population affect the measure of the dispersion of the sample means?

6. What are the characteristics of the sampling distribution of percentages?

Answers to sample exams.

True/False

1. T 2. T 3. T 4. F 5. T 6. F 7. T 8. F 9. F 10. T 11. T 12. F 13. F
14. F 15. T 16. F 17. T 18. F 19. T 20. T 21. T 22. T 23. F 24. F 25. F 26. T
27. T 28. T 29. F 30. T

Multiple Choice

1. e 2. d 3. e 4. d 5. e 6. c 7. c 8. e 9. e 10. d 11. b 12. e 13. d
14. a 15. b 16. c 17. b

Problems and Open Ended

1. Finite population and infinite population.
2. A sample can produce accurate information about the population. Cost, cheaper to gather information from a smaller group. Time, samples can produce accurate information in a shorter period of time. And if it is a destructive test, a sample will not destroy all the members of the population.
3. A sampling distribution of means is the distribution of the arithmetic means of all the possible samples of size n that could be selected from a given population.
4. If the sample size is greater than 30, then the sampling distribution approximates the normal distribution regardless of the kind of distribution of the population.
5. The value of $\sigma_{\overline{x}}$ decreases as the size of the sample, n, increases.
6. $\mu_p = \pi$ and $\sigma_p = \sqrt{\dfrac{\pi(100-\pi)}{n}}$ or $\sigma_p = \sqrt{\dfrac{\pi(100-\pi)}{n}}\sqrt{\dfrac{N-n}{N-1}}$ for a finite population.

T 1. If a sample mean equals 32, then 32 is an estimator of the population mean.

T 2. An estimator is a statistic used to approximate a parameter value.

F 3. An unbiased estimator should always give us the exact approximation of the parameter value.

F 4. An estimator is unbiased if the mean of the sample distribution is equal to the population parameter.

T 5. The standard deviation for a sample is an unbiased estimator of the population standard deviation.

T 6. The precision of an estimate is determined by the amount of sampling error.

F 7. We use the z table to estimate the mean whenever the population values are normally distributed.

T 8. An interval estimate is more acceptable than a point estimate because it gives an idea of the amount of sampling error and thus the probability of being correct can be objectively assessed.

F 9. In practice, a point estimate is usually never computed.

T 10. It is necessary to know the type of probability distribution which a sampling distribution approximates in order to compute an interval estimate.

T 11. The sample percentage is an unbiased estimator of the population percentage.

T 12. A confidence level should be associated with the particular method of estimation and not the specific interval.

T 13. We lose precision in the estimate when a wider confidence interval is desired.

F 14. In practice, when we estimate the population standard deviation, we must also estimate the standard deviation of the sampling distribution.

T 15. The χ^2 distribution can only be used with populations that are normally distributed.

T 16. The sample variance is an unbiased estimator of the population variance.

T 17. Since the shape of the sampling distribution may be determined by the sample size, we should not automatically refer to the z table when computing an interval estimate.

F 18. In interval estimation, the use of either the z table or the t table is determined by the specified confidence level.

T 19. Assume we want a confidence level of 95 percent. If a t value is used, the t value would be larger than a z value with a 95 percent confidence level.

F 20. Since there is sampling error in estimation, the sampling distribution has a standard deviation which is larger than the population standard deviation.

T 21. An increase in sample size reduces the amount of sampling error and thus results in a reduction of the standard deviation of the sampling distribution.

T 22. Sampling error can be controlled by selecting a sample of adequate size.

F 23. Sampling error can be eliminated by selecting a sample of adequate size -- i.e., up to 50% of the population size.

F 24. It is possible to determine the necessary sample size needed for an interval estimate of a population mean without making an estimate of the population standard deviation.

T 25. If you are trying to determine the correct sample size to use to estimate a population percentage, and if you have no idea what that percentage might be, you should assume that it is 50%.

F 26. The χ^2 distribution approaches the normal distribution as n increases.

F 27. A statistical software package can simulate the selection of samples from a population with known parameters, but it can produce a confidence interval for only one sample.

T 28. The sample standard deviation can be computed in such a way that it's a useful estimator of the population standard deviation.

T 29. The shape of a t distribution depends on the sample size.

T 30. A t distribution is used if the sample size is less than 30 and the population standard deviation is unknown.

1. Which of the following is not true about a point estimate?
   (a) A point estimate may result from the use of an unbiased estimator.
   (b) An interval estimate relies on the point estimate to form the range of values.
   (c) A point estimate, like any other estimate, is subject to sampling error.
   (d) A point estimate says nothing about the degree of error in estimation.
X (e) A point estimate is more accurate than an interval estimate.

2. Which of the following is not needed to compute an interval estimate of the population mean?
   (a) Shape of population distribution if $n < 30$.
   (b) Sample size and population size.
   (c) Population standard deviation value.
   (d) Confidence coefficient.
X (e) All of the above are needed.

3. Assume that the average score of a sample of 16 units is 29 with a sample standard deviation of 6. Which, if any, of the following is not true?
   (a) An unbiased estimate of the population mean is 29.
   (b) The sampling distribution approximates a distribution that is flatter than the normal distribution.
   (c) A useful estimate of the population standard deviation is 6.
   (d) It's possible to calculate an interval estimate with the data given if a confidence level is specified.
X (e) All of the above are true.

4. Assume a sample of 25 items from an infinite population has a total score of 350 and the population standard deviation is 5. Which of the following is not true?
   (a) The mean of the sample distribution is 14.
   (b) The degrees of freedom for the t value are 24.
   (c) The standard deviation of the sampling distribution is 1.
X (d) The interval estimate of the population mean will be narrower than the range 13 to 15, if a 95.4% confidence level is used.
   (e) No finite correction factor is needed to estimate the standard deviation of the sampling distribution.

5. If a sample of 36 items has a total value of 3,600 with a population standard deviation of 6, which of the following is not true?
   (a) The mean of the sampling distribution is 100.
   (b) The standard deviation of the sampling distribution is one if we assume that the population is infinite.
   (c) A point estimate of the population mean is 100.
X (d) Without the finite correction factor, the interval estimate of the population mean is the interval 90 to 110 with a confidence coefficient of 95.4 percent.
   (e) The z table is appropriate for this problem.

6. Assume that a sample of 25 items from a normally distributed population has a total value of 2,250. If the population standard deviation is 5 and the finite correction factor is 0.8, which of the following is not true?
   X (a) The finite correction factor is not applicable because the population is normally distributed.
   (b) A point estimate of the population mean is 90.
   (c) The standard deviation of the sampling distribution is .8.
   (d) A 95.4 percent confidence interval for the population mean is 88.4 to 91.6.
   (e) The z distribution is applicable to this problem.

7. Given the following information from a simple random sample of an infinite population, what would be the interval estimate at the 95.4 confidence level?
   $\overline{x} = \$1.20,\ n = 36,$ Pop. std. dev. $= \$0.24$
   X (a) $\$1.12 < \mu < \$1.28$
   (b) $\$1.15 < \mu < \$1.25$
   (c) $\$1.17 < \mu < \$1.26$
   (d) $\$1.18 < \mu < \$1.25$
   (e) $\$1.19 < \mu < \$1.21$

8. A wholesaler of golf clubs wishes to estimate the average amount of money a retailer spends on one order. A simple random sample of 49 order had a mean of \$130 with an estimate of the population standard deviation of \$28. What would be the interval estimate with a 90 percent confidence level?
   (a) $\$121.24 < \mu < \$138.76$
   (b) $\$121.44 < \mu < \$138.56$
   (c) $\$122.16 < \mu < \$137.84$
   X (d) $\$123.42 < \mu < \$136.58$
   (e) $\$124.56 < \mu < \$135.44$

9. Assume a normally distributed infinite population and the following data:
   $\overline{x} = \$150,\ \widehat{\sigma} = 35,\ n = 25,$ confidence level $= 95\%$
   What is the upper limit of the interval estimate?
   (a) \$157.00
   (b) \$161.48
   (c) \$163.72
   (d) \$164.42
   X (e) \$164.45

10. Assume a random sample of 25 items from an infinite normal population produced the following: $\overline{x} = \$200$, $\hat{\sigma} = 20$. What is the lower limit of a 90% confidence interval?
(a) $193.08
X (b) $193.16
(c) $193.19
(d) $194.73
(e) $194.74

11. The owner of a stationery store wants to estimate the average number of pencils sold each day. A random sample of 25 days is selected and the sample mean is 100. The population standard deviation is estimated to be 15. Assuming a 95% confidence level, what is the upper limit of the interval estimate?
(a) 105.12
(b) 105.13
(c) 105.88
X (d) 106.19
(e) 106.88

12. Given the following data, what's the interval estimate of the mean of an infinite normal population? $\overline{x} = 400$, $\sigma = 80$, $n = 25$, confidence level = 90%
(a) $366.98 < \mu < 433.02$
X (b) $372.62 < \mu < 427.38$
(c) $372.67 < \mu < 427.33$
(d) $373.68 < \mu < 426.32$
(e) $382.68 < \mu < 417.32$

13. An assembly line manager wishes to estimate the percentage of defective items produced. A sample of 49 items had 40 percent defective. What would be the approximate upper limit of a 95% confidence interval?
(a) 41%
(b) 46%
(c) 49%
X (d) 54%
(e) 59%

14. A simple random sample of 36 items from a production process had 24 defective items. What would be a 99% confidence interval for estimating the population percentage?
X (a) 46 to 87%
(b) 51 to 83%
(c) 54 to 80%
(d) 60 to 74%
(e) 65 to 69%

15. Assume that the following data are from an infinite population:

$$p = 20\%, \quad n = 49, \quad \text{confidence level} = 99\%.$$

What is the approximate upper limit of the interval estimate?
(a) 26%
(b) 27%
(c) 31%
X (d) 35%
(e) 39%

16. If you don't know the population percentage (and have no idea what it might be) and want to take a sample to estimate its value with a given degree of confidence about the size of the sampling error, you should
(a) forget it -- it can't be done.
X (b) assume that π is 50% and compute the sample size needed.
(c) take a sample of 1,000.
(d) estimate the population size and multiply by 0.2121 to get the sample size.
(e) do none of the above.

17. Any statistic used to estimate a parameter is
(a) unbiased.
(b) biased.
X (c) an estimator.
(d) an interval estimate.
(e) None of the above.

18. "A point estimate is adjusted for _____ _____ to produce an interval estimate." The two words to complete this sentence are:
X (a) sampling error
(b) sampling parameters
(c) population statistics
(d) population samples
(e) None of the above.

19. The degree of probability associated with an interval estimate is known as the
(a) sample variation.
(b) population probability.
(c) degrees of freedom.
X (d) confidence coefficient.
(e) None of the above.

20. Which of the following is not a property of a χ^2 distribution.
    (a) The mean is equal to the degrees of freedom.
    (b) It is not symmetric.
    (c) The degrees of freedom are $n - 1$.
    (d) The distribution values extends from zero indefinitely to the right in a positive direction.
X (e) All of the above are properties of the χ^2 distribution.

1. Explain the difference, if any, between an estimate and an estimator.

2. What are the characteristics of an unbais estimator?

3. What does the confidence coefficient refer to?

4. Which distribution, if any, do we use to form a confidence interval for μ when:
   (a) σ is unknown, $n > 30$, and population distribution is normal?
   (b) σ is known, $n > 30$, and population distribution unknown?
   (c) σ is unknown, $n \leq 30$, and population distribution is normal?
   (d) σ is unknown, $n \leq 30$, and population distribution unknown?

5. What conditions need to be satisfied before we can calculate a confidence interval for the population percentage?

6. How does the χ^2 distribution differ from the normal distribution?

7. Under what conditions must the finite correction factor be used to approximate the standard error of the mean?

8. A warehouse store wants to estimate the average number of laser printer toner cartridges sold each day. A random sample of 36 days is selected and the sample mean is 120. The population standard deviation is estimated to be 15. Construct a 95% confidence interval for the mean.

9. A medical researcher is interested in the proportion of college students who have broken a bone. She takes a sample of 150 college students and finds that 66 have broken a bone. Use this information to form an 99% confidence interval for π.

Answers to sample exams.

True/False

1. T 2. T 3. F 4. F 5. T 6. T 7. F 8. T 9. F 10. T 11. T 12. T 13. T
14. F 15. T 16. T 17. T 18. F 19. T 20. F 21. T 22. T 23. F 24. F 25. T 26. F
27. F 28. T 29. T 30. T

Multiple Choice

1. e 2. e 3. e 4. d 5. d 6. a 7. a 8. d 9. e 10. b 11. d 12. b 13. d
14. a 15. d 16. b 17. c 18. a 19. d 20. e

Problems and Open Ended

1. An estimate is a specific value while an estimator is the statistic that is used to estimate a parameter.
2. An unbiased estimator is one that produces a sampling distribution that has a mean that's equal to the population parameter to be estimated.
3. The probability of correctly including the population parameter being estimated in the interval that is produced.
4. (a) Normal distribution. (b) Normal distribution. (c) t-distribution (d) The confidence interval cannot be computed.
5. $np \geq 500$ AND $n(100 - p) \geq 500$.
6. (1) All χ^2 values are non-negative. (2) A χ^2 distribution is not symmetrical. (3) the mean of a χ^2 distribution is equal to the degrees of freedom.
7. The finite correction factor is to be used when the population from which the sample is finite. However, in practice, most researcher only use the finite correction factor when the sample represent 5% or more of the population.
8. $\overline{x} \pm z\frac{s}{\sqrt{n}}$; $\overline{x} = 120$; $z = 1.96$; $s = 15$; $n = 36$. Thus we have $120 \pm 1.96\left(\frac{15}{\sqrt{36}}\right) =$ 120 ± 4.9 or 115.1 to 124.9 toner cartridges.
9. We have $p = \frac{66}{150} \times (100 \text{ percent}) = 44\%$; $n = 150$; $z = 2.58$; $\widehat{\sigma}_p = \sqrt{\frac{44 \cdot 56}{150}} = 4.05$. Thus we have $p \pm z\,\widehat{\sigma}_p = 44 \pm 2.58(4.05) = 44 \pm 10.45$ or 33.55% to 54.45%.

T 1. In a single sampling, the statistics will most likely not be equal to the parameters.

T 2. With a normal distribution, a knowledge of the population mean and the population standard deviation, we are able to determine the likelihood of particular values for sample statistics.

F 3. With a normal distribution, a knowledge of the population mean and the population standard deviation, we are able to determine the likelihood of particular values for the sample mean by simply looking at the difference between the sample mean and the population mean.

F 4. When statistical results are "significant," the population parameter has been proved to be equal to the null hypothesis.

F 5. The major advantage of hypothesis testing is that a statistician can always make an accurate conclusion because of objectively designed methods.

T 6. In drawing a conclusion about the status of a hypothesis, we not only compute the likelihood of a statistical value based on given parameters but also determine the likelihood of erroneously rejecting the hypothesis.

T 7. The assumption to be tested is the null hypothesis while we reject the null hypothesis in favor of the alternative hypothesis.

T 8. The alternate hypothesis never states that the parameter is a specific value.

T 9. The type of alternate hypothesis depends on the nature of the problem.

F 10. If the null hypothesis is true, the probability of obtaining any difference between sample means and the population mean diminishes as the size of the sample decreases.

F 11. After stating the null and alternate hypotheses, a researcher should specify the level of significance after reviewing the results.

F 12. The level of significance provides a decision rule for the rejection of a null hypothesis. It also represents the probability of being correct when rejecting the null hypothesis.

T 13. When the differences between sample results and the assumed value are converted into normally distributed standard units, we are able to partition this normal curve in critical and rejection regions.

T 14. A hypothesis test can never completely prove that a hypothesis is true.

T 15. The rejection regions of a normal curve are determined by the level of significance and the type of alternate hypothesis.

T 16. Under a two-tailed test, the researcher is not concerned about whether the difference between a sample mean and the hypothesized population mean is positive or negative.

F 17. It is not possible that a null hypothesis could be rejected under a two-tailed test and that the very same null hypothesis and with the same sample statistics we could fail to reject the null hypothesis under a one-tailed test.

F 18. Under a two-tailed test with a level of significance set at 0.05, the risk of erroneous rejection in each tailed is 0.05.

T 19. In contrast to the two-tailed test, the one-tailed test, will reject the null hypothesis on the basis of (1) the size of the difference between the sample mean and the assumed population mean and (2) whether or not the difference is positive or negative.

F 20. In a right-tailed test, the alternate hypothesis states that the true mean is greater than the sample statistic.

T 21. A major disadvantage of any one-tailed test is that we fail to reject the null hypothesis when the actual parameter value may be nowhere close to the assumed value.

F 22. For a right-tailed test where the z distribution is applicable, we can fail to reject the null hypothesis if the test statistic is greater than a z value determined by the level of significance.

T 23. If the population standard deviation is unknown, the z distribution may not be applicable.

F 24. When determining the critical regions under a t distribution, we need to simply use the sample size and the level of significance.

T 25. The t distribution does not apply when we are testing a percentage hypothesis and the sample size is less than 30.

T 26. The costlier it is to mistakenly reject a true hypothesis, the smaller the level of significance should be.

T 27. The level of significance is also the risk of a type I error.

F 28. In some cases, a hypothesis test can prove that the null hypothesis is true.

F 29. The t distribution is used whenever the sample size is less than 30.

F 30. The z distribution can be safely used when the sample standard deviation is known and the sample size is 20 or larger.

1. A production manager fears that a beer bottling machine is, on the average, filling each quart bottle with less than 32 ounces of beer. If this is true, there are possible legal ramifications for the firm. Assuming a hypothesis test is to be performed, which of the following statements is correct?
   (a) The null hypothesis should state that H_0: $\mu = 32$ while the alternate hypothesis would be H_1: $\mu \neq 32$.
   (b) This is a two-tailed test with the alternative hypothesis that H_1: $\mu \leq 32$.
   (c) The null and alternate hypotheses cannot be specified until the level of significance is stated.
   (d) This is a right-tailed test.
X (e) This is a left-tailed test.

2. A college administrator is concerned over stories that the overall average grade for all college students in the U.S. is not 75 (where 75 should represent an average or C grade). If the average grade across the nation is less than 75, it might be an indication that remedial courses should be incorporated into college curricula. If the average grade across the nation is greater than 75, it might be an indication that some action must be taken to restrain instructors from giving grades which are too high. Assuming the administrator took a random sample and wanted to perform a test, which of the following statements is true.
X (a) The null hypothesis should state that H_0: $\mu = 75$ while the alternate hypothesis should be H_1: $\mu \neq 75$.
   (b) It is a two-tailed test where the null hypothesis states H_0: $\mu \leq 75$ and the alternate hypothesis states that H_1: $\mu \geq 75$.
   (c) The null and alternate hypotheses cannot be determined from the given information.
   (d) This is a right-tailed test because the administrator is concerned with only high average grades.
   (e) This is a left-tailed test because the administrator is concerned with only low average grades.

3. A sample of 200 citizens in a metropolitan area is taken and the point estimate of the average annual income per person is $24,600. Before sampling, there was a hypothesis that the average income per person was $25,000. Assuming that a two-tailed test is appropriate with a level of significance of 10 percent, which of the following statements is correct if the population standard deviation is $250?
   (a) Each rejection region under the normal curve has a probability of 0.10.
   (b) No statement has been made that incomes are normally distributed, and so the hypothesis test is invalid.
   (c) The difference between the point estimate and the hypothesized value is -1.75 standard deviates and thus there is a significant difference.
X (d) The test statistic is -22.63, and since it's less than $z = -1.645$ we can reject the null hypothesis.
   (e) None of the above are correct.

4. A brewery wants to test the hypothesis that the typical student at a university drinks an average of 3 cans per week. If the average consumption is less than the hypothesized value, additional advertising will be authorized. Assume that the population standard deviation is known, what other information is needed to test the hypothesis? We need to know...
   (a) only the sample size and whether it is a one- or two- tailed test.
X (b) the sample size, sample mean, and level of significance.
   (c) the level of significance only.
   (d) the alternative hypothesis and the sample size.
   (e) None of the above are correct.

5. You have received a research report to evaluate, and this research report was based on a right-tailed hypothesis test. The conclusion of this report stated that the statistics produced "significant" results at the 0.05 level, which of the following is correct?
   (a) The null hypothesis is most likely correct.
   (b) The true value must be less than the hypothesized value.
   (c) The direction of the rejection region is unknown and thus we can make no conclusions.
   (d) The alternate hypothesis cannot be accepted because it is significantly different from the hypothesized value.
X (e) It is unlikely that the distribution about the hypothesized parameter could have produced a sample statistic so large.

6. A librarian is afraid that less than 200 books are checked out each week. If less than 200 books are checked out, the library officials would consider terminating her job. Assuming the librarian has sampled 10 weeks and the population standard deviation is unknown, which of the following statements is correct if a significance level of 0.05 were used?
   (a) This is a one-tailed test with a 0.025 probability of erroneously rejecting the null hypothesis.
   (b) This is a two-tailed test with the use of the normal distribution to determine the critical region.
X (c) This is a left-tailed test with the use of the t distribution.
   (d) This is a right-tailed test with the use of the normal distribution.
   (e) None of the above are correct.

7. The population mean is assumed to be 50 with $\sigma = 6$. A sample of 36 had $\overline{x} = 47.5$. If a two-tailed test with $\alpha = 0.01$ were conducted, which of the following is correct?
X (a) The test statistic is -2.5 and thus the null hypothesis cannot be rejected.
   (b) The null hypothesis should be accepted because the test statistic is greater than 1.96.
   (c) The test statistic is 2.5 and thus the null hypothesis should be rejected.
   (d) Since $\sigma_{\overline{x}}$ is unknown, a test cannot be made without a point estimate of $\sigma_{\overline{x}}$.
   (e) None of the above are correct.

8. A consumers group decides to test the claim of a car manufacturer that their cars go from 0 to 60 mph in 10 seconds or less. To test this claim, they test drive 20 cars and find the sample mean time was 10.5 seconds with a sample standard deviation of 0.5 seconds. Assuming a significant level of 0.05, which of the following statements is correct?
   (a) This is a two-tailed test with the use of the normal distribution to determine the critical region.
   (b) This is a two-tailed test with the use of the t distribution to determine the critical region.
   (c) This is a one-tailed test with the use of the normal distribution.
X (d) This is a one-tailed test with the use of the t distribution.
   (e) None of the above are correct.

9. The risk of erroneous rejection of the null hypothesis is known as
   (a) the critical region.
   (b) type II error.
   (c) alternative hypothesis.
X (d) the level of significance.
   (e) None of the above are correct.

10. The degrees of freedom in a chi-squared distribution used in a one sample hypothesis test of variance and standard deviation.
   (a) $N - 1$
X (b) $n - 1$
   (c) cannot be determined without knowing the level of significance.
   (d) cannot be determine without knowing if this is a one tailed or two-tailed test.
   (e) None of the above are correct.

11. Psychologist A claimed that 20% of the people in the United States have schizophrenic tendencies. Psychologist B disputed the 20% but had not idea whether a higher of lower percentage was appropriate. Psychologist B randomly sampled 400 people and 72 of then had schizophrenic tendencies. Assuming a significant level of 0.05, which of the following statements is correct?
   (a) This is a left-tailed test with a test statistic of 1.75 and a z value of -1.96.
   (b) This is a right-tailed test with a test statistic of 1.0 and a z value of 1.645.
X (c) This is a two-tailed test with a test statistic of -1.0 and a z value of 1.96 and -1.96.
   (d) This is a right-tailed test with a test statistic of 1.75 and a z value of 1.96.
   (e) This is a left-tailed test with a test statistic of 1.0 and a z value of -1.645.

12. The second step in the hypothesis-testing procedure is to
   (a) determine the test distribution to use.
   (b) state the hypotheses.
X (c) select the level of significance.
   (d) compute the test statistic.
   (e) do none of the above operations.

13. Assume we have a hypothesis that the population percentage is 90% and the alternative hypothesis is $p < 90\%$. If $p = 88\%$, $\sigma_p = 1.2\%$ and $\alpha = 0.05$.
   (a) Reject the null hypothesis because the test statistic is greater than -1.645.
X (b) Reject the null hypothesis because the test statistic is less than -1.645.
   (c) Fail to reject the null hypothesis because the test statistic is less than -1.96.
   (d) Fail to reject the null hypothesis because the test statistic is greater than -1.96.
   (e) None of the above are correct.

14. Which of the following is not a step in hypothesis testing?
   (a) Compute the test statistic.
   (b) Determine the decision rule.
   (c) State the hypotheses.
X (d) Compute the population mean.
   (e) All of the above are hypothesis-testing steps.

15. The term "significant difference" refers to
X (a) the difference between the sample mean and the hypothetical population mean that leads to the rejection of the null hypothesis.
   (b) the difference between the sample standard deviation and the population mean.
   (c) the difference between the decision rule and the test statistic.
   (d) the difference between n and N.
   (e) none of the above.

16. The t distribution is used in hypothesis testing of means when
   (a) $n > 30$.
   (b) $n < 30$, the population is normally distributed, and the population standard deviation is known.
X (c) $n < 30$ and the population standard deviation is unknown.
   (d) the Poisson distribution must be reexamined.
   (e) the binomial distribution must be reexamined.

17. A vending machine manufacturer makes a machine that dispenses 7 ounce of coffee into a cup. The filling mechanism is considered out of control if the standard deviation of the cup weights exceed 0.10 ounces. A sample 12 cups taken during a routine check produce a sample deviation of 0.11 ounces. Assuming that the filling mechanism fills the cups in a normally distributed, which of the following statements is correct?
   (a) This is a two-tailed test with χ^2 distribution with 11 degrees of freedom.
   (b) This is a one-tailed test with H_1: $\sigma < 0.10$ ounces.
   (c) This is a one-tailed test with H_1: $\sigma > 0.11$ ounces.
   (d) This is a one-tailed test with χ^2 distribution with 12 degrees of freedom.
X (e) None of the above are correct.

18. Using the information in problem 17. The test statistic is
    (a) 9.09.
    (b) 12.10.
    (c) 13.20.
X (d) 13.31.
    (e) 14.52.

19. A ski resort operator claims that 25% percent of the people using their slopes are snowboarders. A ticket seller feels that this number is too low. A poll of the 12,500 ticket purchased for one day reveals that 4,200 are snowboards. Assuming a significant level of 0.05, which of the following statements is correct?
    (a) This is a right-tailed test with a test statistic of 22.2 and a z value of 1.96.
X (b) This is a right-tailed test with a test statistic of 22.2 and a z value of 1.645.
    (c) This is a left-tailed test with a test statistic of -20.4 and a z value of -1.96.
    (d) This is a left-tailed test with a test statistic of -20.4 and a z value of -1.645.
    (e) None of the above are correct.

20. In hypothesis testing, the standardized difference between a known statistic and the assumed parameter is called the
    (a) null hypothesis.
    (b) level of significance.
    (c) decision rule.
    (d) acceptance region.
X (e) test statistic.

1. What is a null hypothesis?

2. What are type I errors and type II errors?

3. What is the decision rule and why do we use it?

4. What is a critical value?

5. For each of the following situations, formulate a null and an alternative hypothesis:
   (a) A consumer advocate claims a fruit stands consistently puts less than 5 pounds of apples in their 5 pound bags.
   (b) The service manager of a car dealership claims that they service 145 cars a day.
   (c) A web developer of a company web site claims that at least 50 people visits the web site an hour.
   (d) A tire manufacturer wants to verify that their new tire has a tread life greater than 60,000 miles.
   (e) A city council member claims that 60 percent of the residents favor a new courthouse.
   (f) The chief of police claims that over 50 percent of all crimes in his city are drug and/or alcohol related.
   (g) A meteorologist claims that 80 percent of the people hit by lighting are male.

6. How does a *p*-value procedure differ from the classic hypothesis-testing procedure?

7. How is the classical two-tail test affect by whether σ is known?

8. A vending machine manufacturer makes a machine that dispenses 7 ounce of coffee into a paper cup. The filling mechanism is considered out of control if the standard deviation of the cup weights exceed 0.10 ounces. A sample 12 cups taken during a routine check produce a sample deviation of 0.11 ounces. Assuming that the filling mechanism fills the cups in a normally distributed, what is the test statistic for this hypothesis test?

9. A ski resort operator claims that 25% percent of the people using their slopes are snowboarders. A ticket seller feels that this number is too low. A poll of the 12,500 ticket purchased for one day reveals that 4,200 are snowboards. What is the test statistic for this hypothesis test?

Answers to sample exams.

True/False

1. T 2. T 3. F 4. F 5. F 6. T 7. T 8. T 9. T 10. F 11. F 12. F 13. T
14. T 15. T 16. T 17. F 18. F 19. T 20. F 21. T 22. F 23. T 24. F 25. T 26. T
27. T 28. F 29. F 30. F

Multiple Choice

1. e 2. a 3. d 4. b 5. e 6. c 7. a 8. d 9. d 10. b 11. c 12. c 13. b
14. d 15. a 16. c 17. e 18. d 19. b 20. e

Problems and Open Ended

1. This is the statement of equality. This hypothesis is always assumed to be true even though you often hope to reject it.
2. A type I error is the risk that a true hypothesis will be rejected. While in a type II error we erroneously fail to reject a false hypothesis.
3. A decision rule is a formal statement that clearly states the appropriate conclusion to be reached about the null hypothesis based on the value of the test statistic. We need to state what we will do *before* we calculate the test statistic.
4. The critical value is the start or boundary of the rejection region.
5. (a) H_0: $\mu = 5$; H_1: $\mu < 5$. (b) H_0: $\mu = 145$; H_1: $\mu \neq 145$. (c) H_0: $\mu = 50$; H_1: $\mu > 50$. (d) H_0: $\mu = 60,000$; H_1: $\mu > 60,000$. (e) H_0: $\pi = 60$; H_1: $\pi \neq 60$. (f) H_0: $\pi = 50$; H_1: $\pi > 50$. (g) H_0: $\pi = 80$; H_1: $\pi \neq 80$.
6. Decision rule is to reject H_0 if the p-value $< \alpha$.
7. (1) The correct sampling distribution can no longer be assumed to be normally shaped if $n \geq 30$. (2) We must use an estimated standard error, $\widehat{\sigma}_{\overline{x}}$ instead of $\sigma_{\overline{x}}$.
8. $\chi^2 = \dfrac{(n-1)s^2}{\sigma^2} = \dfrac{11 \cdot (0.11)^2}{(0.10)^2} = \dfrac{0.1331}{0.01} = 13.31.$
9. To calculate the values we have $p = \dfrac{4200}{12500} = 33.6$; $\sigma_p = \sqrt{\dfrac{25 \cdot 75}{12500}} = 0.3873$. So the test statistic is $z = \dfrac{p - \pi_0}{\sigma_p} = \dfrac{33.6 - 25}{0.3873} = 22.2.$

T 1. Two-sample hypothesis testing procedures are not primarily concerned with estimating the absolute values of the parameters.

F 2. A requirement of two-sample hypothesis testing procedures is that the samples be dependent on each other or related in some significant way.

F 3. The null hypothesis in a two-sample test is that the populations differ with respect to some given characteristic.

F 4. Unlike one-sample tests, two-sample tests have only one form for the alternative hypothesis.

T 5. Two-sample tests are based on the sampling distribution of the differences between parameters.

F 6. The sampling distribution of the differences between sample means will have a value of 1 if the null hypothesis is true.

T 7. The values of two-sample means need not be equal even if the population means from which they were selected were equal.

T 8. The standard error of the difference between means is just a standard deviation of the sampling distribution of differences.

T 9. The hypothesis testing procedure with two-sample problems is quite similar to the procedure used with one-sample tests.

T 10. If the null hypothesis is true in two-sample tests, the mean of the appropriate sampling distribution will have a value of zero.

F 11. It would be impossible for the mean of sample 1 to be larger than the mean of sample 2 if population 2 had a larger mean than population 1.

F 12. In computing the standard error of the difference between sample means, it is not necessary to make use of the sample size of the first sample.

T 13. It is possible to make two-sample tests of means even when the population standard deviations are unknown.

F 14. It is not possible to conduct two-sample hypothesis tests of percentages.

F 15. The sampling distribution of the difference between percentages will have a mean of one if the null hypothesis is true.

F 16. When the data come from independent samples an F distribution is used to test the hypothesis about the variance of two populations.

T 17. In a test of differences between percentages, an estimated standard error must always be used.

F 18. The paired t test is used when you have independent samples of the same size taken from normally distributed populations.

T 19. The z distribution is used to test two means when the size of both samples exceed 30 and the samples are taken from independent and normally distributed populations.

F 20. The χ^2 distribution is used to test two population variances if the samples are taken from independent and normally distributed populations.

1. In conducting two-sample hypothesis tests of means,
   (a) we are primarily interested in estimating the absolute values of the population means.
   X (b) we use the data obtained from two independent samples to make relative comparisons between population means.
   (c) the samples must be selected from the same population.
   (d) the testing procedure bears no resemblance to the procedure used in one-sample tests.
   (e) None of the above is appropriate.

2. If there is a statistically significant difference between the means of two populations,
   (a) the null hypothesis will be equal to zero.
   (b) the alternative hypothesis will be equal to zero.
   (c) the alternative hypothesis must be two-tailed.
   (d) only one alternative hypothesis is possible.
   X (e) None of the above is appropriate.

3. If there is a statistically significant difference between the means of two populations,
   X (a) the null hypothesis will be rejected.
   (b) the alternative hypothesis will be rejected.
   (c) the mean of the sampling distribution of differences between sample means will equal zero.
   (d) the standard deviation of the sampling distribution of differences between sample means will equal zero.
   (e) None of the above is appropriate.

4. If the mean of the first sample is greater than the mean of the second sample,
   (a) the null hypothesis must be false.
   (b) the alternative hypothesis must be true.
   (c) the mean of the first population must be greater than the mean of the second population.
   X (d) it is possible for the mean of the sampling distribution of differences to be equal to zero.
   (e) None of the above is appropriate.

5. If the null hypothesis in a two-sample test of means is true,
   (a) the mean of the first population must be one less than the mean of the second population.
   X (b) the mean of the sampling distribution of differences will equal zero.
   (c) the difference between sample means will equal zero.
   (d) the alternative hypothesis must be two-tailed.
   (e) the alternative hypothesis must be one-tailed.

6. The standard error of the difference between means
   (a) is equal to the population standard deviation of population one.
   (b) cannot be computed if the population standard deviations are unknown.
X (c) is dependent on the size of the samples used.
   (d) must equal zero if the null hypothesis is true.
   (e) is equal to the difference between the sample means.

7. In a two-sample hypothesis test of percentages,
   (a) the null hypothesis is that the population percentages are equal to 50%.
   (b) only a single alternative hypothesis format is possible.
   (c) the samples must be selected from the same population.
   (d) the testing procedure bears no resemblance to any other hypothesis testing procedure.
X (e) None of the above is appropriate.

8. Which distribution is used to conduct a hypothesis test about the means when you have dependent samples taken from normally distributed populations?
   (a) The z distribution.
   (b) A F distribution and then a t distribution.
   (c) A F distribution and then a z distribution.
X (d) A t distribution only.
   (e) None of the above distributions are appropriate.

9. The standard error of the difference between percentages
   (a) is not used in the computation of the test statistic.
X (b) is a standard deviation that must be estimated in two-sample hypothesis tests of percentages.
   (c) will equal zero when the null hypothesis is true.
   (d) will not be affected by changes in sample sizes.
   (e) will not be affected by changes in sample percentages.

10. There is a question of whether or not the means of two populations are equal. A sample of 50 items from the first population had a mean of 130 and $\sigma = 10$ while a sample of 36 items from the second population had a mean of 125 and $\sigma = 18$. With a significance level of 0.05, which of the following is correct?
   (a) This is a two-tailed test with z values of ± 1.96 and a test statistic of 1.51. Thus the null hypothesis should be rejected.
X (b) The z values are +1.96 and -1.96. Since the test statistic falls between these values, the null hypothesis cannot be rejected.
   (c) This is a two-tailed test with z values ± 1.64. Since the test statistic is 2.20, the null hypothesis should be rejected.
   (d) This is a two-tailed test with z values of ± 1.64 and a test statistic of 1.51. Thus the null hypothesis should be accepted.
   (e) None of the above are correct.

11. A researcher is interested in determining whether or not the mean of population 1 is less than the mean of population 2. A sample of 90 items from population 1 had a mean of 252 while a sample of 85 items from population 2 had a mean of 256. Assuming $\sigma_1^2 = 180$ and $\sigma_2^2 = 170$, which of the following is true if a test were conducted with a significance level of 0.05?
    (a) This is a left-tailed test with $z = -1.96$ and a test statistic of -1.25.
    (b) This is a two-tailed test with $z = \pm 1.96$ and a test statistic of -125.
    (c) This is a right-tailed test with $z = 1.96$ and a test statistic of -1.
X (d) This is a left-tailed test with $z = -1.64$ and a test statistic of -2.00.
    (e) This is a two-tailed test with $z = \pm 1.64$ and a test statistic of 1.

12. A researcher wants to determine whether or not the percentage in population A is less than the percentage in population B. A sample of 161 items from population A had $p = 20$ while a sample of 189 items from population B had $p = 25$. If a significance level of 0.05 is used, which of the following statements is correct?
    (a) This is a left-tailed test with $z = -1.96$ and the test statistic is -1.12.
X (b) This is a left-tailed test with $z = -1.64$ and the test statistic is -1.12.
    (c) This is a right-tailed test with $z = 1.64$ and the test statistic is -1.12.
    (d) This is a right-tailed test with $z = 1.96$ and the test statistic is -1.12.
    (e) None of the above are true.

13. The mean of the sampling distribution of differences between sample percentages
X (a) is equal to zero if the null hypothesis is true.
    (b) is equal to zero if the alternative hypothesis is true.
    (c) is equal to the percentage of population one.
    (d) is equal to the percentage of population two.
    (e) is equal to the average of the population percentages if the null hypothesis is true.

14. Suppose that we have two samples taken from two independent normally distributed populations where the population standard deviations are unknown. If the sample sizes are larger than 30 then the next steps in the hypothesis test are?
X (a) Use sample standard deviations and the z distribution to conduct the test.
    (b) Use sample standard deviations and a χ^2 distribution to conduct the test.
    (c) Use sample standard deviations and a t distribution to conduct the test.
    (d) Conduct a F test to test the hypothesis that the variances are equal.
    (e) None of the above.

15. In the procedure used to conduct a hypothesis test about variances of two populations, we specify that the sample with the largest variance is designated as sample 1 (the numerator). Why
    (a) So that we only need to conduct a left-tailed test.
X (b) So that we only need to conduct a right-tailed test.
    (c) Because the alternative hypothesis is always $\sigma_1^2 < \sigma_2^2$.
    (d) Because the alternative hypothesis is always $\sigma_1^2 > \sigma_2^2$.
    (e) None of the above are true.

1. What is the purpose of two-sample hypothesis tests?

2. What are the two necessary conditions that must be valid if the testing procedure for equal variances is to produce usable results?

3. Use a two-tailed test to test the equality of population variance at the 0.05 level of significance with the following data: For sample A: $n = 20$, $s^2 = 17.87$ and $\overline{x} = 45.75$. And for sample B: $n = 16$, $s^2 = 28.53$ and $\overline{x} = 40.33$.

4. What are the assumptions that must be satisfied to produce useful results in a paired t test?

5. What are the assumptions that must be satisfied to produce useful results in z test?

6. What are the assumptions that must be satisfied to produce useful results in a t test?

7. What are the assumptions that must be satisfied to produce useful results in a test of two percentages?

8. Two new subcompact cars and tested to see if they get the same miles per gallon. Fifteen *Synglets* were tested and they averaged 43.75 mpg with $s^2 = 2.48$ mpg. While 8 *Gustos* averaged 40.33 mpg with $s^2 = 12.54$ mpg. Assume that the samples are taken from independent populations whose variance are not equal. Test the hypothesis at 0.01 level that these cars do not get the same miles per gallon.

9. Two brands of flood light bulbs are tested of how long they last. The data for Brand A is $n = 32$, $s^2 = 90.21$ hours, and $\overline{x} = 2637.84$ hours. The data for Brand B is $n = 45$, $s^2 = 128.21$ hours, and $\overline{x} = 2602.76$ hours. Test the hypothesis at 0.05 level that the Brand A last longer than Brand B. Assume that the samples are taken from normally distributed independent populations.

Answers to sample exams.

True/False

1. T 2. F 3. F 4. F 5. T 6. F 7. T 8. T 9. T 10.T 11. F 12. F 13. T
14. F 15. F 16. F 17. T 18. F 19. T 20. F

Multiple Choice

1. b 2. e 3. a 4. d 5. b 6. c 7. e 8. d 9. b 10. b 11. d 12. b 13. a
14. a 15. b

Problems and Open Ended

1. In two-sample hypothesis tests we are interested in the relative values of a parameter. That is, is there a statistically significant difference between the parameters of two populations.
2. (1) The data in the two populations sampled must be normally distributed. (2) The data source of the two populations are independent.
3. Step 1: H_0: $\sigma_1^2 = \sigma_2^2$; H_1: $\sigma_1^2 \neq \sigma_2^2$. Step 2: $\alpha = 0.05$. Step 3 and 4: F distribution with 15 df in the numerator and 19 df in the denominator, so $F = 2.62$. Step 5: Reject H_0 in favor of H_1 if $\boldsymbol{F} > 2.62$. Step 6: $\boldsymbol{F} = \dfrac{28.53}{17.87} = 1.60$. Step 7: Fail to reject H_0.
4. The samples are dependent and the populations are normally distributed.
5. Both sample are independent and when both samples are > 30 or when the populations are normally distributed and the standard deviations for both populations are known.
6. Both samples are taken from independent populations that are normally distributed and the samples are small (≤ 30). If the populations variance are not equal, use procedure 3. If the population variances are equal, use procedure 4.
7. The two samples are taken from two independent populations and the samples from each population are sufficiently large, $np \geq 500$ and $n(100 - p) \geq 500$.
8. Step 1: H_0: $\mu_1 = \mu_2$ and H_1:: $\mu_1 \neq \mu_2$. Step 2: $\alpha = 0.01$. Step 3: Since the s populations have equal variances. We use a t distribution and procedure 3. Step 4: The critical t values with 7 df is ± 3.499. Step 5: Reject H_0 in favor of H_1 if $t < -3.499$ or $t > +3.499$. Otherwise, fail to reject H_0. Step 6: Then $t = \dfrac{43.75 - 40.33}{\sqrt{\dfrac{2.48}{15} + \dfrac{12.54}{8}}} = 2.606$.
   Step 7: Since $t = 1.268$, we fail to reject H_0.
9. Step 1: H_0: $\mu_1 = \mu_2$ and H_1: $\mu_1 > \mu_2$. Step 2: $\alpha = 0.05$. Step 3: Procedure 2, z distribution. Step 4: The critical z value is 1.645. Step 5: Reject H_0 in favor of H_1 if $z > 1.645$. Otherwise, fail to reject H_0. Step 6: Then $\boldsymbol{z} = \dfrac{2637.84 - 2602.76}{\sqrt{\dfrac{90.21}{32} + \dfrac{128.21}{45}}} = 14.7$.
   Step 7: Since $\boldsymbol{z} > 1.645$. We reject H_0, Brand A last longer than Brand B.

T 1. ANOVA can be used to determine if there are significant differences between three or more sample means.

T 2. The null hypothesis in analysis of variance (ANOVA) is always that independent samples are drawn from different populations with the same mean.

F 3. Both one and two-tailed tests are possible with analysis of variance when three or more samples are considered.

F 4. The ANOVA technique will tell you exactly which population means differ from the others, and the amount of the difference.

T 5. It is assumed in analysis of variance that the samples are drawn randomly and each sample is independent of the other samples.

T 6. An ANOVA assumption is that the populations under study have distributions which approximate the normal curve.

F 7. An ANOVA assumption is that the populations from which the samples are obtained have different population variances.

F 8. If the null hypothesis is true in an ANOVA test, the population means will not all be equal, but the variances will all be the same.

F 9. The ANOVA technique is appropriate when the population distributions are skewed and when the population variances are unequal.

T 10. In an analysis of variance, two estimates of the population variance are computed on the basis of two independent computational approaches.

F 11. In the ANOVA technique, three computed estimates of the population variance are prepared.

T 12. If the null hypothesis is true, the computed estimates of the population variance should be approximately the same.

T 13. The mean of the variances found within each of the samples is an estimator of the population variance that remains appropriate regardless of the size of the population means.

T 14. The computed F value in ANOVA should ideally be one if the null hypothesis is true.

F 15. If the null hypothesis is true, the result of an ANOVA test will produce a computed F value that must be equal to 1.

F 16. In the ANOVA technique, if the computed F test statistic exceeds the F distribution table value the null hypothesis is accepted.

F 17. The $\hat{\sigma}^2_{\text{between}}$ value computed in ANOVA will yield a reliable estimate of the population variance regardless of whether or not the null hypothesis is true.

F 18. The $\hat{\sigma}^2_{\text{within}}$ value computed in ANOVA will yield a reliable estimate of the population variance only if the null hypothesis is true.

F 19. Some formulas used in ANOVA must change depending on whether or not the sizes of the samples are equal.

T 20. If an ANOVA test is being made at the 0.05 level, the critical value found in the F table is simply the point on the F scale beyond which an F test statistic would be expected to fall only 5 times in 100 if the null hypothesis were true.

F 21. An F distribution is always skewed to the left.

F 22. A single F distribution is used for all combinations of sample number and sample size.

T 23. In using the F distribution, you must determine the degrees of freedom for both the numerator and the denominator of the computed F test statistic.

F 24. In using the F distribution table, the degrees of freedom for the numerator is found by subtracting the number of samples from the total number of items in all samples.

T 25. In ANOVA, if the computed F test statistic is greater than the F value found in the F distribution table, the null hypothesis should be rejected.

T 26. The estimate of the population variance that remains appropriate regardless of any differences between population means is often called the MS error value.

T 27. The estimate of the population variance that is appropriate if (and only if) the population means are equal is often called the MS factor value.

F 28. An ANOVA table shows many of the intermediate values needed to produce a computed F test statistic, but it doesn't actually compute that figure.

T 29. An ANOVA table shows the degrees of freedom needed for a test, but it doesn't give the critical table value of F.

T 30. The 95% confidence intervals produced below the ANOVA table by the Minitab software package give us an idea of the intervals that are likely to include the population means.

1. In the analysis of variance (ANOVA) technique,
   (a) the null hypothesis is that independent samples are drawn from a single population.
   (b) the alternative hypothesis may be left tailed.
   (c) the alternative hypothesis may be right tailed.
X (d) the null hypothesis is always that the different populations have the same mean.
   (e) two of the above are correct.

2. In using ANOVA,
   (a) it is possible to determine exactly which population means differ from the others.
X (b) it is assumed that the samples are independent.
   (c) it is assumed that the population distributions are skewed.
   (d) it is assumed that judgment samples are selected.
   (e) None of the above are appropriate.

3. In making an ANOVA test,
   (a) three estimates of the population variance are computed.
   (b) two dependent computational approaches are used to estimate the population variance.
X (c) one estimated value of the population variance becomes the standard against which the second value can be evaluated.
   (d) two estimates of σ^2 will always yield the same results.
   (e) the value of the two estimates of σ^2 would be exactly the same if the null hypothesis is true.

4. The effects of differences in the means of the populations under study in an ANOVA test should appear in the
X (a) $\widehat{\sigma}^2_{\text{between}}$ value.
   (b) $\widehat{\sigma}^2_{\text{within}}$ value.
   (c) F distributions tables.
   (d) test at the 0.05 level only.
   (e) None of the above is appropriate.

5. In computing $\widehat{\sigma}^2_{\text{within}}$,
   (a) we arrive at a value that is influenced by changes in population mean values.
X (b) we find the arithmetic mean of the estimates of the population variance that have been obtained from each sample.
   (c) only one sample variance is computed.
   (d) we utilize concepts founded on the Central Limit Theorem.
   (e) None of the above is appropriate.

6. In computing $\hat{\sigma}^2_{\text{between}}$,
   (a) we arrive at a value that is not influenced by changes in population mean values.
   (b) we compute the variances of each sample and take the mean of those variances.
   (c) we must arrive at the same value as the one found using the $\hat{\sigma}^2_{\text{within}}$ approach.
   (d) we will usually get a value between zero and one.
X (e) None of the above is appropriate.

7. The difference between the estimates of $\hat{\sigma}^2$ in ANOVA is statistically significant when
   (a) the computed F ratio is between 0 and 1.
   (b) the computed F ratio is between 1 and 2.
   (c) the computed F ratio is between 0 and 2.
X (d) the computed F test statistic > the F table value.
   (e) the computed F test statistic < the F table value.

8. The "grand" mean used in ANOVA is
   (a) the mean of the first two sample means.
   (b) always the mean of all the sample means.
X (c) the mean of all the sample values.
   (d) the mean of all the sample means less the number of degrees of freedom.
   (e) Two of the above.

9. The computed F ratio in ANOVA
   (a) is found by dividing the "between" estimate into the "within" estimate.
   (b) would ideally have a value of zero if the null hypothesis were true.
X (c) must be compared with a "critical value" in a table of F distribution values.
   (d) will be less than the appropriate table F value if the null hypothesis is false.
   (e) None of the above is appropriate.

10. An F distribution
   (a) is skewed to the left in most cases.
   (b) is determined by the number of samples.
   (c) is determined by the number of observations in the samples.
   (d) has only a single critical value.
X (e) Two of the above are appropriate.

11. In using an F distribution table,
    (a) the degrees of freedom for the numerator of the computed F test statistic is one more than the number of samples.
X   (b) the degrees of freedom for the numerator of the computed F test statistic is one less than the number of samples.
    (c) the level of significance of the test is not needed.
    (d) we find a high and low critical value for each situation.
    (e) Two of the above are appropriate.

12. The estimate of the population variance that remains appropriate regardless of any difference between population means is called the
X   (a) MS error.
    (b) MS factor.
    (c) SS factor.
    (d) SS error.
    (e) None of the above.

13. The estimate of the population variance that is appropriate if (and only if) the population means are equal is called the
    (a) SS factor.
    (b) MS error.
    (c) SS error.
    (d) grand mean.
X   (e) None of the above.

14. If the "p" figure at the top right corner of a MINITAB ANOVA table is 0.062 and a test is made at the 0.05 level,
    (a) the null hypothesis can't be determined.
    (b) the alternative hypothesis will be accepted.
X   (c) the null hypothesis will be accepted.
    (d) the alternative hypothesis is that the population means are equal.
    (e) None of the above are correct.

15. The 95% confidence intervals produced at the bottom of a MINITAB ANOVA table show the intervals that are likely to include the
    (a) population variances.
X   (b) population means.
    (c) sample variances.
    (d) pooled standard deviations.
    (e) sample sizes.

1. What is the purpose of a one-way ANOVA test of means?

2. What are assumptions must be true in order to apply ANOVA?

3. What are the null and alternative hypotheses in any ANOVA test?

4. What is the grand mean?

5. Which measure $\hat{\sigma}^2_{\text{within}}$ or $\hat{\sigma}^2_{\text{between}}$ is an estimate of σ^2 if (and only if) the null hypothesis is true?

6. Explain the df entries, and tell how they are computed.

7. Conduct an ANOVA test at 0.05 level if there are 6 samples with a total of 46 items in all of the samples, and if $\hat{\sigma}^2_{\text{between}} = 215.23$ and $\hat{\sigma}^2_{\text{within}} = 72.78$.

Answers to sample exams.

True/False

1. T 2. T 3. F 4. F 5. T 6. T 7. F 8. F 9. F 10. T 11. F 12. T 13. T 14. T 15. F 16. F 17. F 18. F 19. F 20. T 21. F 22. F 23. T 24. F 25. T 26. T 27. T 28. F 29. T 30. T

Multiple Choice

1. d 2. b 3. c 4. a 5. b 6. e 7. d 8. c 9. e 10. b 11. b 12. a 13. e 14. c 15. b.

Problems and Open Ended

1. It is test in which only one factor or variable is considered and we want to compare three or more unknown population means.
2. (1) The population under study must have normal distribution. (2) The samples are drawn randomly and must be independent. (3) The populations from which the samples are obtain all have the same (unknown) variance.
3. H_0: All population means are equal. H_1: Not all population means are equal.
4. The mean of all the values in all the samples.
5. $\widehat{\sigma}^2_{\text{within}}$
6. With k samples and T items total, $\text{df}_{\text{num}} = k - 1$ and $\text{df}_{\text{den}} = T - k$.
7. Step 1: H_0: All population means are equal, and H_1: Not all the population means are equal. Step 2: $\alpha = 0.05$. Steps 3 and 4: $F_{5,40,0.05} = 2.45$. Step 5: Reject H_0 and accept H_1 if $\boldsymbol{F} > 2.45$, otherwise fail to reject H_0. Step 6: $\boldsymbol{F} = \dfrac{\widehat{\sigma}^2_{\text{between}}}{\widehat{\sigma}^2_{\text{within}}} = \dfrac{215.23}{72.78} = 2.96$. Step 7: Since 2.96 is greater than 2.45, we reject the H_0. At least one mean is different.

T 1. Chi-square analysis can be used to test the hypothesis that 3 or more independent samples come from populations that have the same percentage of a given characteristic.

F 2. A chi-square value may have a positive or negative sign.

T 3. In any hypothesis testing situation, we are always concerned with the computation of the value of some test statistic for which we know the appropriate distribution.

T 4. The alternative hypothesis for a contingency table test is that one variable is dependent or related to the other variable.

F 5. The scale of possible chi-square values extends from zero indefinitely to the right or to the left.

F 6. The chi-square probability distribution is a single probability curve.

F 7. The shape of a chi-square distribution is independent of the number of degrees of freedom that exists in a given problem.

F 8. For a significance level of 0.05, the critical chi-square value is always less than the degrees of freedom.

T 9. In a chi-square test, if the observed and expected frequencies are identical, the computed chi-square value will be zero.

F 10. From the standpoint of verifying the null hypothesis, a computed chi-square value of 100 would be ideal.

F 11. It is likely that the observed and expected frequencies will be identical if the null hypothesis is true.

T 12. A computed chi-square value is compared with a table chi-square value to determine if the computed value is significantly above zero.

F 13. Chi-square table values divide the area of acceptance into two equal parts.

T 14. If a computed chi-square value exceeds the appropriate table value, this is cause to reject the null hypothesis.

T 15. There is a separate chi-square distribution for each df value.

T 16. A contingency table shows all the cross-classifications of the variables being studied--i.e., it accounts for all contingencies.

T 17. A convenient formula may be used to compute the expected frequencies for each cell of a contingency table.

T 18. The total of the observed and expected frequencies in a chi-square test must be equal.

T 19. In order to locate the appropriate chi-square table value, you must know the degrees of freedom for the particular problem.

F 20. The degrees of freedom are found in the same way for a k-sample hypothesis test and for a test of goodness-of-fit.

T 21. In a 5×4 contingency table, the df will equal 12.

F 22. The decision rule in a chi-square hypothesis test is to accept the null hypothesis if the computed chi-square value is greater than the appropriate table value.

F 23. A goodness-of-fit test may be used to determine if the fit of a normal distribution and a uniform distribution are equal.

T 24. The purpose of a goodness-of-fit test is to decide if the sample results are consistent with the results that would have been obtained if a random sample had been selected from a population with a known distribution.

T 25. A uniform distribution is one in which the frequencies are equal or uniform.

F 26. The formula for computing the chi-square value in a k-sample hypothesis test differs from the formula used to compute the chi-square value in a goodness-of-fit test.

F 27. In a test to determine if a population distribution is uniform, the df value is one more than the number of classes being considered.

1. In a chi-square distribution,
   (a) the values may be either positive or negative.
X (b) the scale extends from zero indefinitely to the right in a positive direction.
   (c) the mode or peak of the curve will be at df = 3.
   (d) the scale extends from zero in both a positive and negative direction.
   (e) None of the above is appropriate.

2. Which, if any, of the following is not a step in the general chi-square testing procedure:
   (a) Formulate the null and alternative hypotheses.
   (b) Select the level of significance to be used.
   (c) Compute the frequencies that would be expected if the H_0 is true.
   (d) Compute a chi-square value and compare it with a chi-square table value.
X (e) All of the above are chi-square testing steps.

3. In a chi-square test,
   (a) the observed frequencies are computed by using the assumption that the H_0 is true.
   (b) a computed chi-square value must be between zero and five if the H_0 is to be accepted.
X (c) a computed chi-square value of zero would guarantee the acceptance of the null hypothesis.
   (d) the observed and expected frequencies must be equal if we are to accept the null hypothesis.
   (e) the procedure ends with the computation of the chi-square value.

4. In a 4 × 4 contingency table,
   (a) there are 8 cells in the table.
   (b) a convenient formula may be used to compute the observed frequencies.
   (c) the expected frequencies must follow a uniform distribution pattern.
X (d) there will be 9 degrees of freedom.
   (e) None of the above is appropriate.

5. In comparing the computed chi-square value with the appropriate chi-square table value,
X (a) the null hypothesis will be accepted if the computed value is less than the table value.
   (b) the area of acceptance is always to the right of the computed value.
   (c) the area of acceptance is always to the left of the computed value;
   (d) the null hypothesis will be rejected if the computed value is less than the table value.
   (e) None of the above is appropriate.

6. The appropriate chi-square table value in a given test:
   (a) Falls two df to the right of the acceptance region.
   X (b) Separates the area of acceptance from the area of rejection.
   (c) Is generally to the left of the area of acceptance.
   (d) Must be less than the computed value if the null hypothesis is to be accepted.
   (e) None of the above is appropriate.

7. Which of the following statements is true:
   (a) The computed chi-square value must be zero if the H_0 is true.
   (b) If the computed chi-square value less than the table value, then we must reject H_0.
   (c) A 6 × 6 contingency table will have 36 degrees of freedom.
   X (d) If a chi-square distribution has 7 degrees of freedom, the mean of this distribution will be 7.
   (e) If the observed and expected frequencies are identical, the null hypothesis must be rejected.

8. A chi-square goodness of fit test
   (a) uses the formula $(r-1)(x-1)$ to determine the degrees of freedom to use.
   (b) is limited to testing the goodness of fit between sample data and a uniform distribution.
   X (c) may be used to evaluate the goodness of fit between a population under study and a normal distribution.
   (d) does not follow the same basic procedures employed by the k-sample test.
   (e) None of the above is appropriate.

9. If an experiment with a hypothesized uniform distribution has k outcomes, then each cell should contain
   X (a) $100/k$ percent of all the sample outcomes.
   (b) $k/100$ percent of all the sample outcomes.
   (c) k percent of all the sample outcomes.
   (d) 10 percent of all the sample outcomes.
   (e) None of the above.

1. Explain the purpose of a goodness-of-fit test and what are the assumptions?

2. What is the objective of a contingency table test, or test of independence?

3. A toy store receives a shipment of 3 new action toys from a TV cartoon. She then observes how many of each character is brought in the first 100 sold..

Action Toy	Number Purchased
Blue Action Figure	32
Red Action Figure	43
Green Action Figure	25

   Test the hypothesis at the 0.05 level that the 3 action toys are equally preferred.

4. Even if the null hypothesis is true, why is it unlikely the computed test statistic will be zero?

Answers to sample exams.

True/False

1.	2.	3.	4.	5.	6.	7.	8.	9.	10.	11.	12.	13.
14.	15.	16.	17.	18.	19.	20.	21.	22.	23.	24.	25.	26.
27.	28.	29.	30.	31.	32.	33.	34.	35.	36.	37.	38.	39.
40.	1.	2.	3.	4.	5.	6.	7.	8.	9.	10.		
3.	3.	3.	3.	3.	3.							

Multiple Choice

1.	2.	3.	4.	5.	6.	7.	8.	9.	10.	11.	12.	13.
14.	15.	16.	17.	18.	19.	20.	21.	22.	23.	24.	25.	26.
27.	28.	29.	30.	31.	32.	33.	34.	35.	36.	37.	38.	39.
40.	1.	2.	3.	4.	5.	6.	7.	8.	9.	10.		
3.	3.	3.	3.	3.	3.							

Problems and Open Ended

1. The purpose is to test if an observed distribution is representative of a hypothesized population distribution. The assumptions are (1) The sample is a simple random sample and frequency counts are obtained for the possible k cells; (2) the expected frequency for each of the k cells is at least 5.
2. The objective is to see if two variables are independent. That is, to see if the occurrence (or non-occurrence) of one does not affect the probability of occurrence of the other.
3. Step 1: H_0: The action toys are equally preferred. H_1: The action toys are not equally preferred. Step 2: $\alpha = 0.05$. Step 3: χ^2. Step 4: The critical χ^2 value with 2 df is 5.99. Step 5: Reject H_0 in favor of H_1 if$\chi^2 > 5.99$. Otherwise, fail to reject H_0. Step 6: $\chi^2 = \sum\left[\frac{(O-E)^2}{E}\right] = 0.05 + 2.83 + 2.07 = 4.95$. Step 7: Since a $\chi^2 < 5.99$, we fail to reject H_0.
4. For the test statistics to be exactly zero, all observation must match all expected outcomes which is highly unlikely.

F 1. The term "relationship" in statistics refers to the cause and effect among variables.

T 2. An objective of regression analysis is to develop an equation which estimates the value of one variable given the value of one or more other variables.

T 3. It is possible that two series of numbers may vary together and there is no cause-and-effect between the series.

F 4. If there is spurious correlation, the r^2 can never be over 0.90.

T 5. One benefit of a scatter diagram is that we can visually determine if there is a useful relationship between the variables before computing any equation.

T 6. If the correlation coefficient for two series of values is a negative value, then the regression line must always slope downward.

F 7. If the coefficient of determination for two series of values is a positive value, then the regression equation must always have a positive slope.

T 8. If we know the value of the y-intercept, the mean of the dependent variable and the mean of the independent variable, then we have enough information to determine the slope of a simple regression line.

F 9. If the correlation coefficient for two series of numbers is zero, then the simple regression line is a vertical line.

F 10. If there is no statistical relationship between two series of numbers, the intercept term of the regression line must be zero.

T 11. If the coefficient of determination for two series of numbers is exactly one, we know that the regression line has passed through all the data points, but we do not know from just this coefficient whether the slope of the line is positive or negative.

F 12. If the slope of a simple regression line is zero, then the correlation coefficient is zero. Likewise, if the slope of the line is one, then the correlation coefficient is one.

F 13. If the correlation coefficient is 0.9, the percentage of variation in the independent variable explained by the dependent variable is 0.81.

F 14. One of the disadvantages with simple regression analysis is that we can only obtain interval estimates and not point estimates.

T 15. There is an inverse relationship between $s_{y.x}$ and r^2.

T 16. Two scatter diagrams may yield the exact same regression equations and still have unequal r^2.

T 17. Assume we have two series of numbers which we will call Series A, and we have another different set of series of numbers which we will call Series B. It is possible that the intercept terms and slope terms of the regression line for A and B are equal. However, at the same time it is possible that the r^2 for A and B are unequal.

F 18. Regression analysis, unlike other statistical aids for forecasting, eliminates the need for subjective estimates in decision making because of the certainty of the relationships between the dependent and independent variables.

F 19. Although we should not always assume that statistics show cause and effect between variables, there is an exception to this rule when $r^2 = 1$.

F 20. In regression analysis, the dependent and independent variables must be the same type of units, e.g., both must be dollars or both must be dozens and so forth.

F 21. When the slope of a regression equation is zero, it always means that no line can fit the data.

T 22. When three or more variables are considered, a multiple regression study may be needed.

F 23. The first person to perform a regression analysis was Thomas Watt.

T 24. If the slope of the population regression line (B) has a value of zero, there's no relationship between the variables.

T 25. A sample might suggest that a relationship exists between variables when B = 0.

F 26. The null hypothesis in the t-test for slope is that B = 1.

F 27. The standard error of estimate isn't needed to conduct a t-test for slope.

T 28. It's possible to compute a confidence interval estimate of the value of B.

F 29. Statistical software packages are useful only when there are two variables to analyze.

T 30. ANOVA concepts can be used to test for the presence of slope in the population regression line.

F 31. The results of an ANOVA test differ from those of a t-test for slope in a simple linear regression situation.

F 32. Predicting the range for the average value of y given x produces a larger interval than predicting the range for a specific value of y given x.

T 33. When small samples are used in preparing confidence interval estimates of y given x, the standard error of estimate value must be adjusted.

T 34. The coefficient of determination is equal to SSR/SST.

1. The slope of a regression line is 32 while the intercept is 20. What would be the value of the dependent variable if the independent variable was 6?
   (a) −88
   (b) −0.44
   (c) 58
   (d) 152
   X (e) 212

2. A regression line has a slope of 16 and an intercept of 4, the mean of the independent variable is 8, what is the mean the dependent variable?
   (a) -2
   (b) 34
   (c) 48
   (d) 84
   X (e) 132

3. If you have been told that the slope of a regression line is 8, how should you interpret the 8?
   (a) It is the value of the dependent variable when the independent variable is zero.
   (b) For every 8 units of increase in the independent variable the dependent variable will increase by one.
   X (c) For every unit of increase in the independent variable, the dependent variable will increase by 8 units.
   (d) It means that the regression equation is a horizontal line 8 units above the axis representing the independent variable.
   (e) It is impossible to interpret without further information.

4. Assume that the independent variable may take on any value from 0 to infinity and that no matter what value the independent variable has, the dependent variable is always 45. Which of the following statements is correct?
   (a) No regression line can be computed.
   (b) The intercept term is zero.
   (c) The slope term and intercept term are zero.
   (d) The slope term is 45 and the intercept is zero.
   X (e) The intercept is 45 and the slope is zero.

5. Assume that we wanted to perform a regression analysis with test scores on history exams as a dependent variable and effort in studying as an independent variable. Past records show that no matter what amount of effort is devoted to studying, i.e., whether 0 or 100 percent effort, the test score is always 62.
   Which of the following statements is correct?
   (a) No regression line can be computed.
   (b) The intercept term is zero.
   (c) The slope and intercept are zero.
   (d) The slope is 62 and the intercept is zero.
   X (e) The intercept is 62 and the slope is zero.

6. There's no relationship between two variables in a simple linear regression analysis if

X (a) $b = 0$.
(b) $r = +1.00$.
(c) $r = -1.00$.
(d) the relationship is negative.
(e) the relationship is positive.

7. The alternative hypothesis in a t-test for slope is that
(a) $B = 0$.
(b) $B = 1$.
(c) $B = 2$.
(d) $B = 3$.
X (e) None of the above.

8. A basketball coach wants to obtain a regression equation where the team's game scores are the dependent values and the points per game by Don Dribble are the independent values.

Team's Game Scores	Dribble's Points
75	16
110	8
90	10
85	14

Which of the following equations is correct?
(a) $\hat{y} = -24.63 + 5.16x$
(b) $\hat{y} = 112.61 - 1.27x$
X (c) $\hat{y} = 135.00 - 3.75x$
(d) $\hat{y} = 93.28 + 1.24x$
(e) $\hat{y} = -14.16 + 4.6x$

9. Assume we have the following data about the time for the 220 yard dash and age of runners:

Time in Seconds	Age
52	43
34	31
29	28
25	26

If a sprinter is 29 years old, what would be her expected time for the 220 yard dash?
(a) 33.5
(b) 32.6
(c) 31.4
X (d) 30.3
(e) 29.8

10. Assume we have the following past data:

y	x
160	8
125	4
145	7
130	5

What is the standard error of estimate if y is the dependent variable?
(a) 13.75
(b) 10.11
(c) 9.43
X (d) 3.71
(e) 1.93

11. A store manager wishes to forecast the number of air conditioners sold during the summer. The independent variable is the average temperature for the summer. He obtained the following past data:

Air Conditioners Sold	Average Summer Temperature
55	72
64	78
58	74
63	76

Assume that the average temperature for the coming summer is expected to be 75 what would a regression equation forecast for the summer air conditioner sales be?
(a) 54
(b) 58
X (c) 60
(d) 62
(e) 64

12. The manager of a large bakery wishes to perform a regression analysis with the number of pretzels produced per minute by an employee as the dependent variable and the number of year of experience as a pretzel maker. He obtained the following data:

Number of Pretzels per Minute	Years Experience
55	3
35	6
45	2
25	1

If someone is looking for a pretzel job and says he has 4 years experience, how many pretzels per minute can we expect him to make?
(a) 33.2
X (b) 40.7
(c) 45.43
(d) 48.24
(e) 55.6

13. Which of the following is not needed to calculate a small-sample 95% confidence interval for the value of B?
    (a) The value of the sample slope.
    (b) A t distribution value.
X (c) The unit of measure of the xy variable.
    (d) The standard error of estimate.
    (e) All of the above are needed.

14. Assume that with two series of numbers the explained variation is 640 and the unexplained variation is 2, 560. What is the coefficient of correlation?
    (a) 0.49
    (b) 0.25
    (c) 0.20
X (d) 0.45
    (e) 0.05

15. With the following information, what is the coefficient of determination?

Dependent Variable	Independent Variable
45	8
48	10
32	16
35	14

    (a) 0.64
    (b) 0.78
    (c) −0.88
    (d) 0.92
X (e) 0.85

16. Which, if any, of the following statements is true?
    (a) As r increases, the coefficient of determination decreases.
X (b) As r decreases, the standard error of estimate increases.
    (c) When r is zero, the standard error of estimate is one.
    (d) When the slope of the regression line is negative, the value of r will be between zero and +1.00.
    (e) None of the above are true.

17. Which of the following is not an underlying assumption that must apply if simple regression analysis techniques are to be used correctly?
    (a) There's a linear relationship between x and y, and fixed values of A and B, in the population.
X (b) For each possible value of x, there's a single y value in the population scatter diagram.
    (c) A homoscedastic condition applies.
    (d) The values of a and b are estimates of A and B.
    (e) All of the above assumptions apply.

18. An ANOVA procedure
    - (a) may be used to test for the presence of a y-intercept in the population regression line.
    - (b) may be used with simple (but not multiple) regression situations.
    - X (c) gives the same results as a t-test for slope in a simple linear regression situation.
    - (d) eliminates the need for hypotheses in making tests.
    - (e) has no place in regression analysis.

19. Calculating an interval estimate for the average value r of y given x produces a range that's
    - (a) the same for both large and small samples.
    - (b) doesn't require the use of the standard error of estimate when the samples are small.
    - (c) doesn't require the use of the standard error of estimate when the samples are large.
    - X (d) produces a range that's smaller than one calculated to include the specific value of y given x.
    - (e) can't be done when the samples are small.

20. The total variation in a scatter diagram
    - (a) equals SST + SSR.
    - X (b) equals SSR + SSE.
    - (c) is the sum of the deviations of the x values about their mean.
    - (d) is the sum of the deviations of the y values about their mean.
    - (e) is the numerator of the coefficient of determination.

1. What is the purpose of linear regression analysis and how does it differ from correlation analysis?

2. What are the properties that the name "least squares" refer to?

3. What is the standard error of estimate?

4. Discuss the roles of the dependent (response) variable and the independent (explanatory) variable(s).

5. Below is the data from a college entrance exam and the students first-year GPA. The data and some calculations are in the table that follows.

Student	$x =$ Exam	$y =$ College GPA	xy	x^2	y^2
1	16	2.64	42.24	256	6.9696
2	22	3.40	74.8	484	11.56
3	23	3.15	72.45	529	9.9225
4	20	3.13	62.6	400	9.7969
5	18	2.77	49.86	324	7.6729
6	10	2.20	22	100	4.84
7	22	3.56	78.32	484	12.6736
8	22	3.31	72.82	484	10.9561
9	13	2.44	31.72	169	5.9536
10	20	3.17	63.4	400	10.0489
11	18	3.14	56.52	324	9.8596
12	20	3.03	60.6	400	9.1809
TOTAL	224	35.94	687.33	4354	109.4346

Additionally, the following were previously calculated:

$$\overline{x} = \frac{224}{12} = 18.6667, \qquad \widehat{y} = .1.2166 + 0.09527x, \qquad s_{y.x} = 0.151.$$

Test to decide if there is a relationship between the entrance exam and the GPA. Use a level of significance of 0.01.

6. Explain the difference between the range of "average value of y given x" and "a specific value of y given x".

7. What is the purpose of a ANOVA test in a multiple linear regression situation.

Answers to sample exams.

True/False

1. F 2. T 3. T 4. F 5. T 6. T 7. F 8. T 9. F 10. F 11. T 12. F 13. F
14. F 15. T 16. T 17. T 18. F 19. F 20. F 21. F 22. T 23. F 24. T 25. T 26. F
27. F 28. T 29. F 30. T 31. F 32. F 33. T 34. T

Multiple Choice

1. e 2. e 3. c 4. e 5. e 6. a 7. e 8. c 9. d 10. d 11. c 12. b 13. c
14. d 15. e 16. b 17. b 18. c 19. d 20. e

Problems and Open Ended

1. The purpose is to describe the relationship between two variables. While the correlation analysis is used to measure the strength of the relationship.
2. "Least squares" refers to the *minimum* or *least* value of the *square* of $(y - \widehat{y})$. The minimum of $\Sigma(y - \widehat{y})^2$.
3. The standard error of estimate is a standard deviation that measures the scatter of the observed values around regression (or estimate) line.
4. The dependent or response variable is the one that is to be estimated. An independent or explanatory variable is used to explain variations in the dependent variable.
5. Step 1: H_0: $B = 0$; H_1: $B \neq 0$. Step 2: $\alpha = 0.01$. Step 3 and 4: t distribution with 10 df, so t value is 3.169. Step 5: Reject H_0 if the test statistic > -3.169 or $< +3.169$.

   Step 6: $s_b = \dfrac{s_{y.x}}{\sqrt{\Sigma(x^2) - \dfrac{(\Sigma x)^2}{n}}} = \dfrac{0.151}{\sqrt{4354 - \dfrac{(224)^2}{12}}} \approx 0.01149$. So

   $t = \dfrac{b}{s_b} = \dfrac{0.09527}{0.01149} \approx 8.29$. Step 7: Reject H_0, there is a significant relationship between the entrance exam and GPA.
6. The average value of y given x is the interval in which a mean of small sample of values of a specific value of y given x.
7. It is to test if there is a meaningful relationship between the y variable and all of the predicting variables.

F 1. Nonparametric statistics are sometimes referred to as distribution-full statistics.

F 2. Nonparametric statistics require that assumptions be made about the shape of a distribution.

F 3. Chi-square methods are parametric in nature.

T 4. Nonparametric statistical methods should be used when rank or ordinal data are used.

T 5. Ordinal data only tell us if one item is higher than, lower than, or equal to another item.

F 6. Nonparametric methods cannot be used when nominal data are available.

F 7. The sign test cannot be used when you have paired ordinal measurements.

T 8. The sign test is based on the signs, negative or positive, of the differences between pairs of ranked data.

F 9. The first step in the sign test is to determine the sign of differences between paired observations.

T 10. Both one- and two-tailed sign tests may be conducted.

F 11. In a sign test, the symbol "r" is given to the larger of the positive or negative signs that are accumulated.

T 12. The binomial probability distribution may be used in a sign test to determine the likelihood of observed sample results.

T 13. The decision rule to follow in small-sample sign tests is to accept the null hypothesis if the level of significance is equal to or greater than the probability of the sample results.

T 14. If the sample size is large (over 30) in a sign test, the normal approximation to the binomial distribution may be used.

F 15. The Wilcoxon signed rank test is used when only the direction of the differences within pairs is known.

T 16. The Wilcoxon signed rank test incorporates the size of the differences in ranks.

F 17. In a Wilcoxon signed rank test, if the computed T value is greater than the table T value, the null hypothesis should be rejected.

F 18. In using the sign test or the Wilcoxon signed rank test, it is necessary that the paired data be drawn from two independent samples.

F 19. To use the Mann-Whitney test, the paired data must be drawn from one sample or two closely related samples.

T 20. The Mann-Whitney test is often referred to as the U test, since a statistic called U is computed for testing the null hypothesis.

F 21. The first step in the Mann-Whitney test is to rank the data.

F 22. In the Mann-Whitney test, the value which is selected for U in hypothesis testing is the greater of two values computed with different formulas.

T 23. The decision rule in a Mann-Whitney test is to reject the null hypothesis of the computed U value is equal to or less than the appropriate value in the U table.

T 24. The purpose of a runs test is to determine if randomness exists or if there is an underlying pattern in a sequence of sample data.

F 25. The first step in a runs test is to count the number of runs.

F 26. The decision rule in a runs test is to reject the null hypothesis if the sample r value is greater than an appropriate table r value.

T 27. The Spearman rank correlation coefficient is a measure of the closeness of association between two ordinal variables.

T 28. If the Spearman correlation measure is zero, there is no correlation between the variables.

T 29. The Spearman correlation measure cannot exceed a value of $+1.00$ or be less than -1.00.

F 30. It is not possible to conduct a hypothesis test using the Spearman correlation measure.

1. Nonparametric methods should not be used in which of the following situations:
   (a) When the sample size is very small and when no assumption can be made about the shape of the population distribution.
   (b) When ranked data are used.
   (c) When ordinal data are used.
   (d) When nominal data are used.
X (e) Nonparametric methods may be used in all of the above situations.

2. In a sign test, n is the
   (a) number of observations.
   (b) number of positive signs.
   (c) number of negative signs.
   (d) smaller of the number positive or negative signs.
X (e) None of the above is true.

3. The sign test
X (a) uses paired ordinal measurements taken from the same subjects or matched subjects.
   (b) makes use of the size of the differences between paired groups.
   (c) must have a one-tailed alternative hypothesis.
   (d) must have a two-tailed alternative hypothesis.
   (e) has as its first step the determination of the differences between paired observations.

4. The sign test procedure with small samples
   (a) is identical with the procedure used for large samples.
   (b) uses the normal approximation to the binomial distribution.
X (c) results in the acceptance of the null hypothesis if the level of significance is equal to or less than the probability of the sample results.
   (d) results in the computation of a T statistic.
   (e) results in the computation of a U statistic.

5. The Wilcoxon signed rank test
   (a) focuses solely upon the direction of the differences within pairs.
   (b) is named for Bruce Wilcoxon, an early 18th century statistician.
   (c) requires the use of two independent samples.
   (d) results in the computation of a U statistic.
X (e) None of the above are appropriate.

6. In the Mann-Whitney test,
   (a) Paired data must be drawn from one sample or two closely related samples.
   (b) a T statistic is computed.
   (c) if the computed T value is less than the table T value, the null hypothesis is rejected.
X (d) a statistic called U is computed for testing the null hypothesis.
   (e) None of the above is appropriate.

7. In the Wilcoxon signed rank test,
   X (a) both the size and the sign of the differences between the paired data are computed.
   (b) the plus and minus signs of the ranked data can never be ignored.
   (c) the positive and negative ranks are added, and the larger of these two sums is designated as the computed T value.
   (d) the computed T value cannot be greater than 10.
   (e) the null hypothesis is accepted if the computed T value is equal to or less than the table T value.

8. In the Mann-Whitney test,
   (a) the first step is to compute the U statistic.
   (b) the U statistic is computed using four different formulas.
   (c) two computed values are obtained for U, and the larger one is used in hypothesis testing.
   (d) the null hypothesis is rejected if the computed U value is greater than the table U value.
   X (e) None of the above is appropriate.

9. In a runs test for randomness,
   X (a) the purpose is to determine if randomness exists or if there is an underlying pattern in a sequence of sample data.
   (b) the nature of a sequence pattern can be determined.
   (c) we can determine whether a pattern is in an upward or downward direction.
   (d) the first step is to count the number of runs.
   (e) None of the above is appropriate.

10. The Spearman rank correlation coefficient
   (a) is a measure of the degree of relationship using the actual values of x and y.
   (b) has a value of zero when the correlation is perfect.
   X (c) has a value of +1.00 when there is perfect positive correlation.
   (d) requires the ranking of only one of the variables.
   (e) is a parametric measure.

11. Which of the following statements is true:
   (a) Distribution-free statistics is a name given to analysis of variance techniques.
   (b) Nonparametric statistics require that assumptions be made about the shape of a distribution.
   X (c) The Mann-Whitney test should be used when differences between paired data are under study, and when the data are drawn from two independent samples.
   (d) In the case of a single sample with sequential data, a Wilcoxon signed rank test may be conducted.
   (e) The Spearman rank correlation coefficient uses a single sample with sequential data.

1. What are nonparametric statistical methods used for?

2. What are sign tests used for?

3. What is the null hypothesis for a sign test?

4. If the difference in the paired data used in a sign test yield 5 positive, 15 negative, and 9 tie or zero values:
   (a) What are the values for n and for r?
   (b) In a right-tailed test at the 0.05 level, should the null hypothesis be accepted or rejected?

5. How does Wilcoxon signed ranked test differ from the other sign test?

6. How do the Mann-Whitney test differ from the Wilcoxon signed ranked test?

7. How are the ANOVA tests and the Kruskal-Wallis Procedure assumptions similar and different?

8. What are the hypotheses in a runs test?

Answers to sample exams.

True/False

1. F 2. F 3. F 4. T 5. T 6. F 7. F 8. T 9. F 10. T 11. F 12. T 13. T
14. T 15. F 16. T 17. F 18. F 19. F 20. T 21. F 22. F 23. T 24. T 25. F 26. F
27. T 28. T 29. T 30. F

Multiple Choice

1. e 2. e 3. a 4. c 5. e 6. d 7. a 8. e 9. a 10. c 11. c

Problems and Open Ended

1. Nonparametric statistical methods are those that produce test results for decision making that don't require that restrictive assumption be made about the shape of a distribution. Also when the ranked or nominal data are used.
2. Sign test are used to test for difference in population means when "before-and-after" situations involving dependent samples are considered.
3. The null hypothesis states that $p = 0.5$, meaning that 50 percent of the signs are positive. That is, a positive sign and a negative sign are equally likely to occur.
4. (a) Since $n =$ the total number of positive and negative scores, $n = 5 + 15 = 20$ (zeros and ties are ignored). There are 5 positive scores (which is less than the 15 negative scores) so $r = 5$.
   (b) The sum of the binomial probabilities is 0.0207. At the 0.05 level, we reject the H_0 since $0.0207 < 0.05$.
5. Wilcoxon signed rank test uses magnitude as well as direction to see if there are true differences between pairs of data drawn from random and dependent samples.
6. The data for a Mann-Whitney test come from independent samples while the data for the Wilcoxon signed ranked test come from dependent or related samples.
7. They both assume equal variances but the ANOVA test require that the populations have *normal distributions* but the Kruskal-Wallis Procedure does not require this.
8. The H_0 used in a runs test is that there is randomness in the sequential data. The H_1 is that there is an underlying pattern in the data.